Advances in Intelligent Systems and Computing

Volume 332

Series editor

Janusz Kacprzyk, Polish Academy of Sciences, Warsaw, Poland
e-mail: kacprzyk@ibspan.waw.pl

About this Series

The series "Advances in Intelligent Systems and Computing" contains publications on theory, applications, and design methods of Intelligent Systems and Intelligent Computing. Virtually all disciplines such as engineering, natural sciences, computer and information science, ICT, economics, business, e-commerce, environment, healthcare, life science are covered. The list of topics spans all the areas of modern intelligent systems and computing.

The publications within "Advances in Intelligent Systems and Computing" are primarily textbooks and proceedings of important conferences, symposia and congresses. They cover significant recent developments in the field, both of a foundational and applicable character. An important characteristic feature of the series is the short publication time and world-wide distribution. This permits a rapid and broad dissemination of research results.

More information about this series at http://www.springer.com/series/11156

Ishwar K. Sethi
Editor

Computational Vision and Robotics

Proceedings of ICCVR 2014

Editor
Ishwar K. Sethi
Department of Computer Science
and Engineering
Oakland University
Rochester, MI
USA

ISSN 2194-5357 ISSN 2194-5365 (electronic)
Advances in Intelligent Systems and Computing
ISBN 978-81-322-2195-1 ISBN 978-81-322-2196-8 (eBook)
DOI 10.1007/978-81-322-2196-8

Library of Congress Control Number: 2014956559

Springer New Delhi Heidelberg New York Dordrecht London

Printed on acid-free paper

Springer (India) Pvt. Ltd. is part of Springer Science+Business Media (www.springer.com)

Preface

Computer Vision and Robotics is one of the most challenging areas in the twenty-first century. Its applications range from agriculture to medicine, domestic to defense, deep-sea to space exploration, and manufacturing and transportation. Today's technologies demand intelligent machines, enabling newer applications in various domains and services.

We are pleased to bring the Proceedings of the International Conference on Computational Vision and Robotics (ICCVR-2014). The aim of the conference was to provide a forum for researchers and scholars to present their research results along with the demonstration of new systems and techniques in the area of Computer Vision and Robotics. Papers from several different areas, such as early vision, shape/range/motion analysis, signal/image processing, image matching, pattern/face recognition, 3D vision/perception, and cognition and robotics were covered and discussed during the conference.

Like every year, this time also we received more than 80 papers for ICCVR-2014 held at Bhubaneswar during August 9–10, 2014. This volume covers chapters from various areas of Computational Vision such as Image and Video Coding and Analysis, Image Watermarking, Noise Reduction and Cancellation, Block Matching and Motion Estimation, Tracking of Deformable Object using Steerable Pyramid Wavelet Transformation, Medical Image Fusion, CT and MRI Image Fusion based on Stationary Wavelet Transform. The book also covers articles from applications of soft computing techniques such as Target Searching and Tracking using Particle Swarm Optimization, PSO-based Functional Artificial Neural Network, etc. The book also covers articles from the areas of Robotics such as Solar Power Robot Vehicle, Multi Robot Area Exploration, Intelligent Driving System based on Video Sequencing, Emotion Recognition using MLP Network, Identifying the Unstructured Environment.

I am sure that the participants must have shared a good amount of knowledge during the conference. I must thank all the participants of ICCVR-2014 and hope that next year also they will attend the conference during the same time. I wish all success in their academic endeavors.

Srikanta Patnaik
Programme Chair

Conference Organizing Committee

General Chair

Prof. Ishwar K. Sethi, Department of Computer Science and Engineering, Oakland University, USA

Programme Chair

Prof. Srikanta Patnaik, Professor, School of Computer Science and Engineering, Siksha 'O' Anusandhan University and Chairman, IIMT, Bhubaneswar, Odisha, India

Program Committee

Dr. Xiaolong Li, Indiana State University, USA
Dr. Yeon Mo Yang, Kumoh University, Korea
Dr. Sugam Sharma, Iowa State University, USA
Dr. Debiao He, Wuhan University, China
Dr. Yaser I. Jararweh, Jordan University of Science and Technology, Jordan
Dr. Nadia Nouali-Taboudjemat, Research Centre on Scientific and Technical Information, Algeria
Prof. Doo Heon Song, Yong-in SongDam College, South Korea
Prof. Baojiang Zhong, Soochow University, China
Prof. Nitaigour Mahalik, California Sate University, Fresno, CA, USA
Prof. Guangzhi Qu, Oakland University, USA
Prof. Peng-Yeng Yin, National Chi Nan University, Taiwan
Prof. Yadavalli Sarma, University of Pretoria, South Africa

Dr. Akshya Kumar Swain, The University of Auckland, New Zealand
Prof. Rabi Mahapatra, Texas A&M University, USA
Prof. Reza Langari, Texas A&M University, USA
Prof. Kazumi Nakamatsu, University of Hyogo, Japan
Prof. Zhihua cui, Taiyuan University of Science and Technology, China
Dr. Arturo de la Escalera Hueso, Intelligent Systems Lab, Spain
Dr. Ashish Khare, University of Allahabad, Allahabad, India
Dr. Mahesh Chandra, Birla Institute of Technology Mesra, Ranchi, India
Prof. Sushanta Panigrahi, IIMT, Bhubaneswar, India
Prof. Hemanta Palo, SOA University, India

Conference Coordinator

Ms. Soma Mitra, Interscience Research Network, Bhubaneswar, Odisha, India

Contents

About the Editor

Ishwar K. Sethi is currently Professor in the Department of Computer Science and Engineering at Oakland University in Rochester, Michigan, where he served as chair of the department from 1999 to 2010. From 1982 to 1999, he was with the Department of Computer Science at Wayne State University, Detroit, Michigan. Before that, he was a faculty member at Indian Institute of Technology, Kharagpur, India, where he received his Ph.D. degree in 1978.

His current research interests include data mining, pattern classification, multimedia information indexing and retrieval, and social media analysis. He has graduated 24 doctoral students and has authored or coauthored over 160 journal and conference articles. He has served on the editorial boards of several prominent journals including IEEE Transactions on Pattern Analysis and Machine Intelligence, and IEEE Multimedia. He was elected IEEE Fellow in 2001 for his contributions to artificial neural networks and statistical pattern recognition.

Tracking of Deformable Object in Complex Video Using Steerable Pyramid Wavelet Transform

Om Prakash, Manish Khare, Rajneesh Kumar Srivastava and Ashish Khare

Abstract Tracking of deformable moving object in complex video is one of the challenging problems in computer vision. Random motion, varying size and shape of the object, and varying background and lighting conditions make the tracking problem difficult. Many researchers have tried to handle this problem using spatial domain-based methods, but those methods are not able to handle movement of object properly in case of varying size, shape, and background of object. In this paper, we have proposed a tracking algorithm for deformable video object. The proposed method is based on the computation of steerable pyramid wavelet transform. Coefficients at different level of decomposition and velocity of object are used to predict object location in consecutive frames of video. The approximate shift-invariance and self-reversibility properties of steerable pyramid wavelet transform are useful for tracking of object in wavelet domain. The translation in object is well handled by shift-invariance property, while self-reversibility property yields to make it useful to handle object boundaries. Experimental results of the proposed method and its comparison with other state-of-the-art methods show the improved performance of the proposed method.

Keywords Object tracking · Shift-invariance · Video processing · Steerable pyramid wavelet transform

O. Prakash (✉)
Centre of Computer Education, Institute of Professional Studies,
University of Allahabad, Allahabad, India
e-mail: au.omprakash@gmail.com

M. Khare · R.K. Srivastava · A. Khare
Department of Electronics and Communication, University of Allahabad, Allahabad, India
e-mail: mkharejk@gmail.com

R.K. Srivastava
e-mail: rkumarsau@gmail.com

A. Khare
e-mail: ashishkhare@hotmail.com

I.K. Sethi (ed.), *Computational Vision and Robotics*, Advances in Intelligent Systems and Computing 332, DOI 10.1007/978-81-322-2196-8_1

1 Introduction

Tracking of object with motions in a video is one of the challenging problems in computer vision [1]. Video object tracking is to locate the object in consecutive frames with time. Object tracking problem is important because its applications are found in wide range of areas, pedestrian and car tracking on the roads, extraction of highlights in a sport video, tracking human faces in surveillance [2], etc. Variation in size, shape, and background, occlusion of the object, cluttered background, and interaction between multiple objects are the major causes involved in the tracking of object in a video. Object tracking algorithms are mainly categorized into region based [3], contour based [4], and feature based [5]. Several parameters such as object size, color, shape, and velocity are involved in region-based tracking, and hence, region-based tracking methods are computationally costly.

In general, feature-based object tracking methods use color histogram processing in spatial domain [3], but these methods are not suitable to handle full occlusion of object and difficult in implementation. In subsequent years, some improvements in tracking results have been made by the use of Bayesian and particle filters in color histogram [6] and kernel-based tracking [7]. These methods yield a balance between computational cost and accuracy of tracking. Object tracking methods based on single feature are used to perform operations on point, shape, or contour in spatial domain [2, 5]. Tracking based on other methods deals with frequency value processing of pixels known as transform-domain processing [8]. Recent trend is to use wavelet transform-based methods.

Although discrete wavelet transform (DWT) provides a fast local sparse and decorrelated multiresolution of video frames, it has the major disadvantages of shift variance and poor directional selectivity [9]. Several complex wavelet transform-based methods have been proposed [10]. Steerable pyramid wavelet transform has been used in our proposed method. The shift invariance helps in accurate tracking, while self-reversibility property is useful to retain object's boundary which is much desired for object tracking.

The proposed algorithm of object tracking uses energy of steerable pyramid wavelet transform coefficients as a feature of object. The use of a single feature makes the tracking fast and accurate. For tracking, in addition to the matching of energy of steerable pyramid wavelet transform, the object movement is predicted on the basis of Newton's equations of motion. For performance evaluation, the proposed method is compared with other state-of-the-art methods: particle filter-based tracking [6], Kernel filter-based tracking [7], and Bayesian filter-based tracking [11].

The rest of the paper is organized as follows: Sect. 2 presents an overview of the steerable pyramid wavelet transform; in Sect. 3, the proposed method of object tracking is given. The experimental results and conclusions of the work have been presented in Sects. 4 and 5, respectively.

2 Steerable Pyramid Wavelet Transform

Unlike most discrete wavelet transforms, the steerable pyramid wavelet transform [12] is a linear multiscale, multiorientation image decomposition method, which is self-reversible, and translation and orientation invariant. The decomposition of input video frame is performed resulting in low-pass subband and high-pass subband using steerable filter H_0 and L_0, respectively. Then, low-pass subband is decomposed into a set of different oriented band-pass subband components $B_1, B_2, B_3, \ldots, B_N$ and a low-pass subband component L_1 for the next level. This low-pass subband of next level L_1 is further decomposed into high-pass and low-pass subbands. The lower low-pass subband is subsampled by a factor of 2 along the x and y directions.

Overall structure of steerable pyramid from analysis to synthesis is shown in Fig. 1.

2.1 *Usefulness of Steerable Pyramid Wavelet Transform in Object Tracking*

The steerable pyramid wavelet transform is a multiscale, multiorientation, self-reversible image transform. Shift-invariance property of steerable pyramid wavelet transform is useful to ensure the movements in the object at different directions, and locally rotating the image transform allows better edge preservation, keeping the shape of object which is much desired in object tracking in case of deformable object. Moreover, the steerable pyramid guarantees perfect reconstruction due to the self-reversibility characteristic. Therefore, it could efficiently facilitate object tracking in a video with varying size, shape, and background of the object.

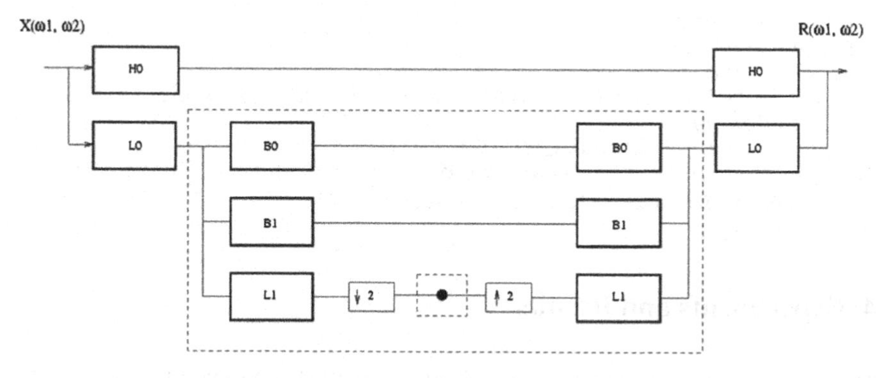

Fig. 1 Block diagram of steerable pyramid transform

3 The Proposed Method

In the proposed algorithm, it is assumed that the frame rate is adequate and the size of the object should not change between adjacent frames. It is also assumed that the object will never acquire a velocity so that it can escape to its neighborhood. Complete algorithm is given as below:

Algorithm: *Object Tracking*

1: Initialize frame_num = 1
2: Draw a bounding_box around the object with centroid (c1, c2) and compute energy of its Steerable pyramid wavelet coefficients ξ as

$$\xi = \sum_{(i,j)\in bounding\_box} \left|wcoeff_{i,j}\right|^2$$

// $wcoeff_{i,j}$ are the Steerable pyramid Wavelet Transform
// coefficients at $(i, j)^{th}$ point
3: for frame_num 2 to end_frame
4: Compute the Steerable pyramid wavelet coefficients of the frame, $wcoeff_{i,j}$
5: Initialize Search_region=32 (in pixels)
6: if frame_num > 4
7: Predict the centroid (c1, c2) of the current frame using centroids of previous four frames and basic Newton's equations of motion.
8: endif
9: for i = -search_region to + search_region do
10: for j = -search_region to + search_region do
11: c1_new = c1+i;
12: c2_new =c2+j;
13: Update bounding_box with centroid (c1_new, c2_new)
14: compute the difference ofenergy of steerable pyramid wavelet coefficients of bounding_box, with ξ, say $D_{i,j}$
15: end for
16: end for
17: Select min{$D_{i,j}$} and its index, say (index_x, index_y)
18: c1 = c1+index_x; c2= c2+index_y;
19: update the bounding_box in the current frame with centroid (c1,c2)and its energy ξ
20: end for

4 Experiments and Results

The algorithm described in Sect. 3 is implemented using MATLAB, and experiments are performed on several video clips of varying object sizes, background, illumination changes, occluded object, ceased object, etc. The visual results for one representative video clip 'soccer video' are shown in Fig. 2. The main challenge of this video is that it has very small object size. For performance evaluation, the proposed method is compared with other state-of-the-art methods: particle filter-based tracking [6], Kernel filter-based tracking [7], and Bayesian filter-based tracking [11].

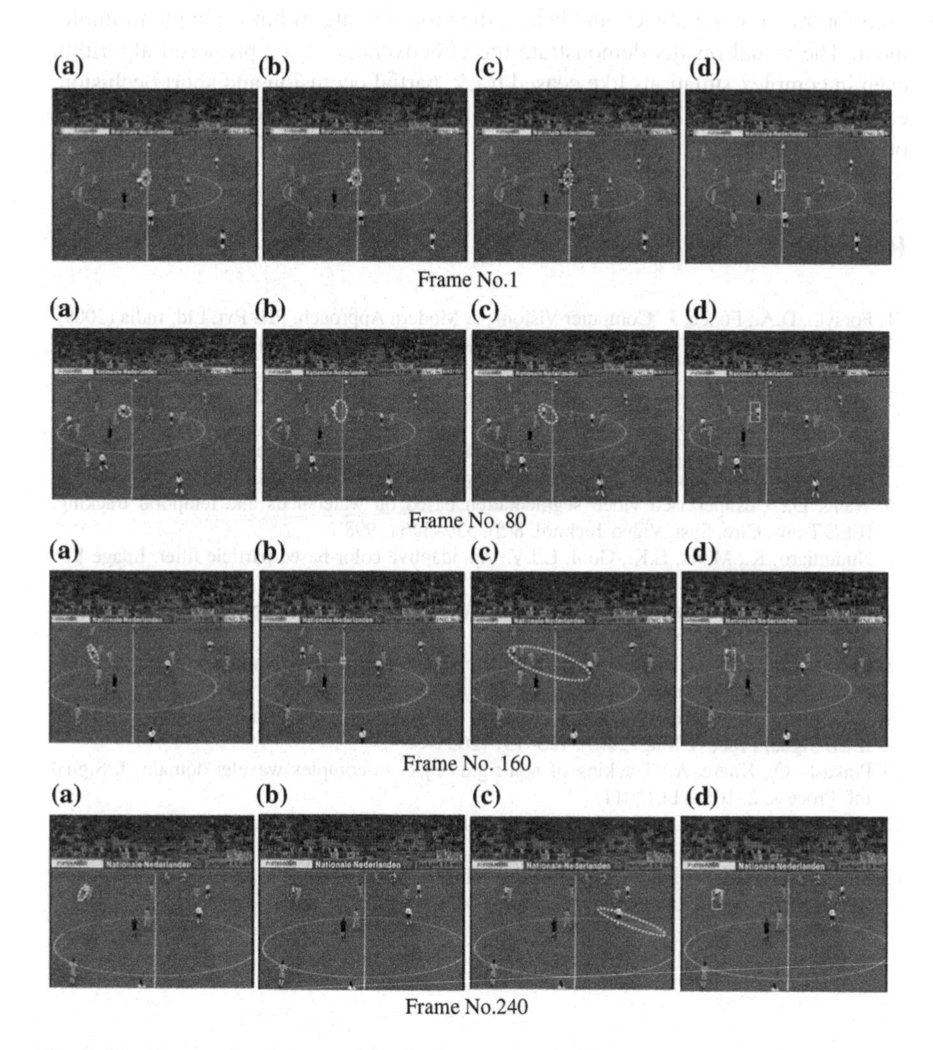

Fig. 2 Tracking results of soccer video for frame no. 1–300 in steps of 80 frames by **a** particle filter-based method [6]. **b** Kernel filter-based method [7]. **c** Bayesian filter-based method [11]. **d** The proposed method

5 Conclusions

In the present work, we have exploited the properties of steerable pyramid wavelet transform which are suitable for tracking of deformable object in complex video. A single-parameter energy of steerable pyramid wavelet transform coefficients is used for tracking the object, and hence, the proposed algorithm is simple to implement. The visual results demonstrate the effectiveness of the proposed algorithm even in complex situations like ceased track, partial occlusion and short occlusion, etc. The experimental results show that the proposed method outperforms over other state-of-the-art methods.

References

1. Forsyth, D.A., Ponce, J.: Computer Vision—A Modern Approach. PHI Pvt. Ltd, India (2009)
2. Yilmaz, A., Javed, O., Shah, M.: Object tracking: a survey. ACM Comput. Surv. **38**(4), (2006)
3. Liu, T.L., Chen, T.: Real-time tracking using trust region methods. IEEE Trans. Pattern Anal. Mach. Intell. **26**(3), 397–402 (2004)
4. Erdem, C.E., Tekalp, A.M., Sankur, B.: Video object tracking with feedback of performance measures. IEEE Trans. Circ. Syst. Video Technol. **13**(4), 310–324 (2003)
5. Wang, D.: Unsupervised video segmentation based on watersheds and temporal tracking. IEEE Trans. Circ. Syst. Video Technol. **8**(5), 539–546 (1998)
6. Nummiaro, K., Meier, E.K., Gool, L.J.V.: An adaptive color-based particle filter. Image Vis. Comput. **21**(1), 99–110 (2003)
7. Comaniciu, D., Ramesh, V., Meer, P.: Kernel-based object tracking. IEEE Trans. Pattern Anal. Mach. Intell. **25**(5), 564–577 (2003)
8. Cheng, F.H., Chen, Y.L.: Real time multiple objects tracking and identification based on discrete wavelet transform. Pattern Recogn. **39**(6), 1126–1139 (2006)
9. Salesnick, L.W., Baraniuk, R.G., Kingsbury, N.: The dual-tree complex wavelet transform. IEEE Signal Process. Mag. **22**(6), 123–151 (2005)
10. Prakash, O., Khare, A.: Tracking of non-rigid object in complex wavelet domain. J. Signal Inf. Process. **2**, 105–111 (2011)
11. Zivkovic, Z., Cemgil, A.T., Krose, B.: Approximate Bayesian Methods for Kernel-based Object Tracking. Comput. Vis. Image Underst. **113**(6), 743–749 (2009)
12. Simoncelli, E.P., Freeman, W.T.: The steerable pyramid: a flexible architecture for multi-scale derivative computation. In: Proceedings of ICIP, pp. 444–447 (1995)

Use of Different Features for Emotion Recognition Using MLP Network

H.K. Palo, Mihir Narayana Mohanty and Mahesh Chandra

Abstract Emotion recognition of human being is one of the major challenges in modern complicated world of political and criminal scenario. In this paper, an attempt is taken to recognise two classes of speech emotions as high arousal like angry and surprise and low arousal like sad and bore. Linear prediction coefficients (LPC), linear prediction cepstral coefficient (LPCC), Mel frequency cepstral coefficient (MFCC) and perceptual linear prediction (PLP) features are used for emotion recognition using multilayer perception (MLP).Various emotional speech features are extracted from audio channel using above-mentioned features to be used in training and testing. Two hundred utterances from ten subjects were collected based on four emotion categories. One hundred and seventy-five and twenty-five utterances have been considered for training and testing purpose.

Keywords Emotion recognition · MFCC · LPC · PLP · NN · MLP · Radial basis function

1 Introduction

Emotion recognition is a current research field due to its wide range of applications and complex task. It is difficult and challenging to analyse the emotion from speech. Emotion is a medium of expression of one's perspective or his mental

H.K. Palo (✉) · M.N. Mohanty
Siksha 'O' Anusandhan University, Bhubaneswar, India
e-mail: hemantapalo@soauniversity.ac.in

M.N. Mohanty
e-mail: mihirmohanty@soauniversity.ac.in

M. Chandra
Birla Institute of Technology, Ranchi, India
e-mail: shrotriya69@rediffmail.com

I.K. Sethi (ed.), *Computational Vision and Robotics*, Advances in Intelligent Systems and Computing 332, DOI 10.1007/978-81-322-2196-8_2

state to others. Some of the emotions include neutral, anger, surprise, fear, happiness, boredom, disgust, and sadness and can be used as input for human–computer interaction system for efficient recognition. Importance of automatically recognising emotions in human speech has grown with increasing role of spoken language interfaces in this field to make it more efficient. It can also be used for vehicle board system where information of mental state of the driver may be provided to initiate his/her safety though image processing approaches [1]. In automatic remote call centres, it is also used to timely detect customer's emotion.

The most commonly used acoustic features in the literature are LPC features, prosody features like pitch, intensity and speaking rate. Although it seems easy for a human to detect the emotional classes of an audio signal, researchers have shown average score of identifying different emotional classes such as neutral, surprise, happiness, sadness and anger. Emotion recognition is one of the fundamental aspects to build man–machine environment that provides theoretical and experimental basis of the right choice of emotional signal for understanding and expression of emotion. Emotional expressions are continuous because the expression varies smoothly as the expression is changed. The variability of expression can be represented as amplitude, frequency and other parameters. But the emotional state is important in communication between humans and has to be recognised properly.

The paper is organised as follows. Section 1 introduces the importance of this work; Sect. 2 represents the related literature. The proposed method has been explained in Sect. 3. Section 4 discusses the result, and finally Sect. 5 concludes the work.

2 Related Literature

The progress made in the field of emotion recognition from speech signal by various researchers so far is briefed in this section. Voice detection using various statistical methods was described by the authors [2] in their paper. The concept of negative, non-negative emotions from a call centre application was emphasised using a combination of acoustic and language features [3]. A review on various methods of emotion speech detection, features and resources available were elaborately explained in the paper [4, 5]. A tutorial review on linear prediction in speech was explored by the author [6] in his paper, and the algorithm behind the representation of speech signal by LP analysis was suitably explained in [6, 7]. Spectral features such as Mel frequency cepstral coefficient (MFCC) was explained completely in [8, 9]. The concept of linear prediction cepstral coefficient (LPCC) and neural network classifier has been the main focus in this paper [10]. Perceptual linear prediction features of speech signals with their superiority over LPC features are suitably proved with experimental results and algorithms by the authors in [11]. The idea about various conventional classifiers including neural network can be found in the paper of authors [12], while speech emotion recognition by using combinations of C5.0, neural network (NN), and support vector machines (SVM) classification methods are emphasised in [13].

3 Proposed Method for Recognition

Two of the major components of an emotional speech recognition system are feature extraction and classification.

3.1 Feature Extraction

Features represent the characteristics of a human vocal tract and hearing system. As it is a complex system, efficient feature extraction is a challenging task in emotion recognition system. Extracting suitable features is one of the main aspects of the emotion recognition system. Linear prediction coefficients (LPCs) [6, 7] are one of the most used features for both speech and emotional recognition. Basic idea behind the LPC model is that given speech sample at time n, $s(n)$ can be approximated as a linear combination of the past p speech samples. A LP model can be represented mathematically

$$e(n) = s(n) - \sum_{k=1}^{p} a_k s(n-k) \tag{1}$$

The error signal $e(n)$ is the difference between the input speech and the estimated speech. The filter coefficients a_k are called the LP (linear prediction) coefficients.

One of the most widely used prosodic features for speech emotion is MFCC [8, 9] which outperformed LPC in classification of speech and emotions due to use of Mel frequency which is linear at low frequency and logarithmic at high frequency to suit the human hearing system.

The Mel scale is represented by the following equation

$$\mathrm{Mel}(f) = 2595 \log_{10}\left(1 + \frac{f}{700}\right) \tag{2}$$

where f is the frequency of the signal.

Linear prediction cepstral coefficients (LPCC) [10] use all the steps of LPC. The LPC coefficients are converted into cepstral coefficients using the following algorithm.

$$\mathrm{LPCC} = \mathrm{LPC}_i + \sum_{k=1}^{i-1}\left(\frac{k-i}{i}\right)\mathrm{LPCC}_{i-k}\mathrm{LPC}_k \tag{3}$$

PLP [11] uses three concepts from the psychophysics of hearing to derive an estimate of the auditory spectrum: (1) the critical-bands spectral resolution, (2) the equal-loudness curve and (3) the intensity-loudness power law. The auditory spectrum is then approximated by an autoregressive all-pole model. A fifth-order all-pole model is effective in suppressing speaker-dependent details of the auditory spectrum. In comparison with conventional LP analysis, PLP analysis is more consistent with human hearing.

The spectrum $P(\omega)$ of the original emotional speech signal is wrapped along its frequency axis ω into the Bark frequency Ω by

$$\Omega(\omega) = 6\ln\left[\frac{\omega}{1200\pi} + \left[\left(\frac{\omega}{1200\pi}\right)^2 + 1\right]^{0.5}\right] \quad (4)$$

where ω is the angular frequency in rad/s.

3.2 Emotion Classification

In this paper, multilayer perception (MLP) [12, 13] classifier is used and the results with different features are compared.

The structure of MLP for three layers is shown in Fig. 1. The three layers are input layer, hidden layer and output layer. Let each layer has its own index variable, '*k*' for output nodes, '*j*' for hidden nodes and '*i*' for input nodes. The input vector is propagated through a weight layer *V*. The output of *j*th hidden node is given by,

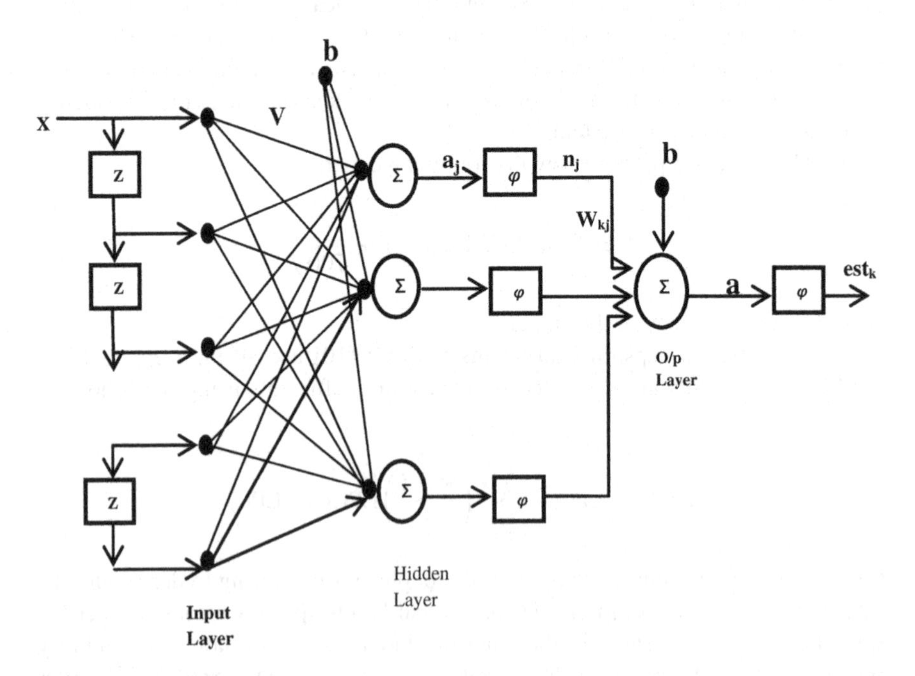

Fig. 1 Structure of MLP

$$n_j = \varphi\big(a_j(t)\big) \tag{5}$$

$$\text{where } a_j(t) = \sum_i x_i(t)v_{ji} + b_j \tag{6}$$

and a_j is output of jth hidden node before activation. x_i is the input value at ith node, b_j is the bias for jth hidden node, and φ is the activation function.

The output of the MLP network is determined by a set of output weights, W, and is computed as

$$\text{est}_k(t) = \varphi(a_k(t)) \tag{7}$$

$$a_k(t) = \sum_j n_j(t)w_{kj} + b_k \tag{8}$$

where η is the learning rate parameter of the back-propagation algorithm.

where est_k is the final estimated output of kth output node. The learning algorithm used in training the weights is back-propagation. In this algorithm, the correction to the synaptic weight is proportional to the negative gradient of the cost function with respect to that synaptic weight and is given as

$$\Delta W = -\eta\frac{\partial\xi}{\partial w} \tag{9}$$

where η is the learning rate parameter of the back-propagation algorithm.

4 Result and Discussion

The database has been prepared for four emotions for a group of 10 subjects for a sentence '*who is in the temple*'. The emotions include boredom, angry, sad and surprise. Database of children in the age group of six to thirteen were selected, including four boys and six girls. Duration of database is 1.5–4 s. The database was recorded using Nokia mobile and converted to wav file using format factory software at 8 kHz sampling frequency with 8 bits. From the Figs. 2, 3, 4 and 5, it was observed that high-arousal emotions like surprise and angry emotions have higher magnitudes than low-arousal emotions, like sad and boredom. Surprise emotions have highest magnitude, while bore emotions have lowest magnitude among all emotions.

As shown in the Table 1. The classification rate of MLP using MFCC feature vectors for the two classes of emotions was found to be highest (80 %) when all the four emotions angry, surprise, sad and bore are taken together. The reorganisation accuracy increases when one of the low-arousal emotions is compared with both the high-arousal emotions. Classification rate is lowest in case of LPC feature

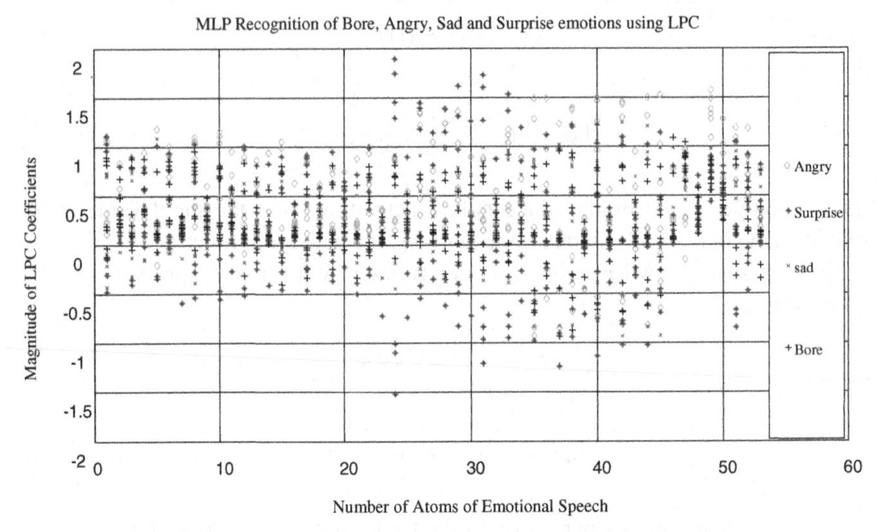

Fig. 2 Recognition of Bore, Angry, Sad and Surprise emotions using LPC

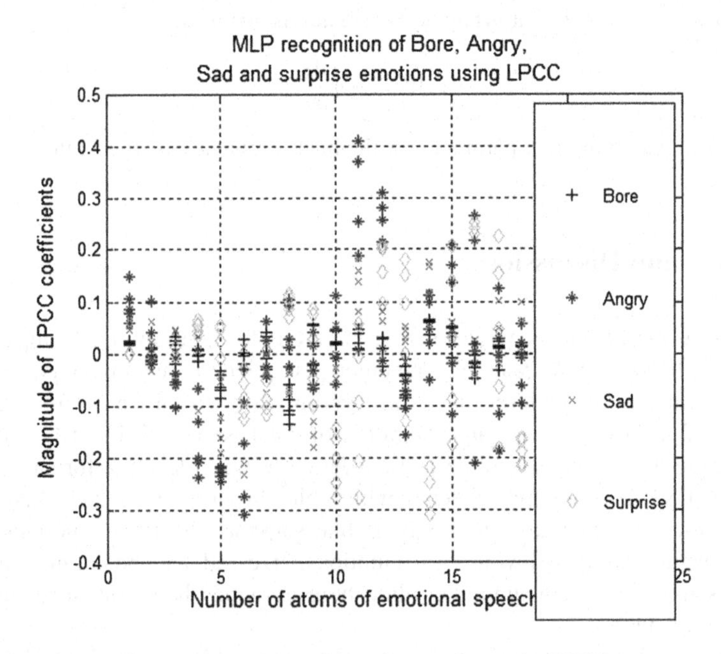

Fig. 3 Recognition of Bore, Angry, Sad and Surprise emotions using LPCC

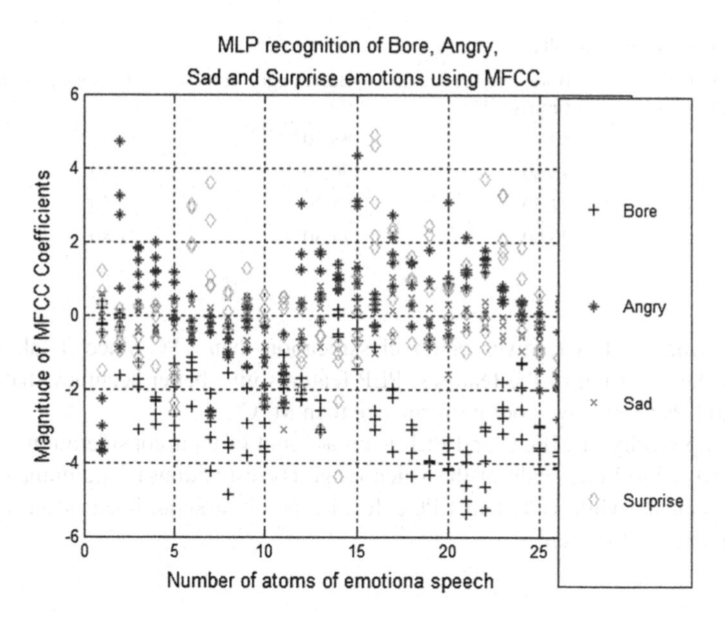

Fig. 4 Recognition of Bore, Angry, Sad and Surprise emotions using MFCC

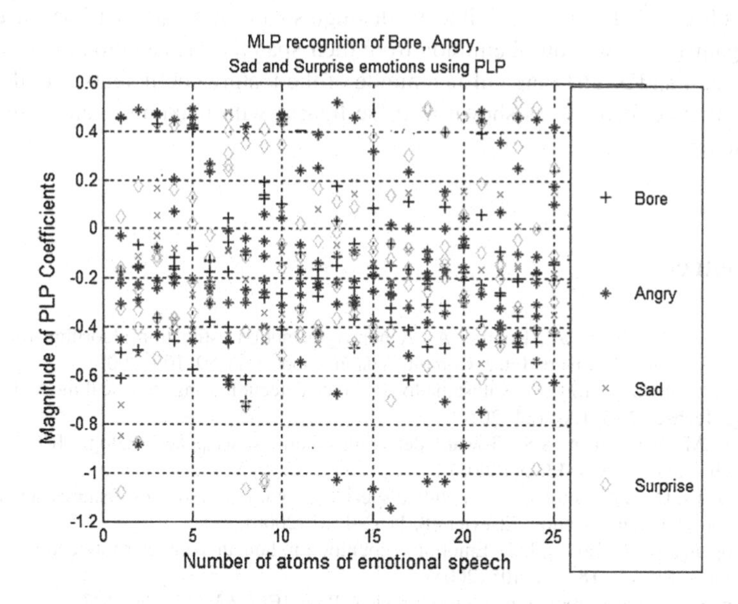

Fig. 5 Recognition of Bore, Angry, Sad and Surprise emotions using PLP

Table 1 Classification results

Feature extraction technique	Bore, angry, sad and surprise (%)	Bore, angry, surprise (%)	Bore, sad, surprise (%)
MFCC	80	83.20	81.40
LPC	48.60	56.20	52.70
LPCC	54.50	65.20	62.50
PLP	70.00	74.30	71.80

vectors, whereas LPCC gives better classification than LPC since it takes into account the cepstrum of the features. PLP features give better accuracy than both LPC and LPCC but gave poor performance than MFCC.

The superiority of MFCC and PLP is on account of their consideration of both linear and logarithmic scale of the voice range corresponding to the human hearing mechanism, while LPC and LPCC feature purely assume linear scale for the entire duration of speech.

5 Conclusion

It was observed that it is possible to distinguish the high-arousal speech emotions against the low-arousal emotion from their spatial representation as shown in Figs. 2, 3, 4 and 5. The range of magnitude of the feature coefficients can identify the emotions effectively as shown in these figures with maximum and minimum dispersions.

References

1. Mohanty, M., Mishra, A., Routray A.: A non-rigid motion estimation algorithm for yawn detection in human drivers. Int. J. Comput. Vision Robot. **1**(1), 89–109 (2009)
2. Mohanty, M.N., Routray, A., Kabisatpathy, P.: Voice detection using statistical method. Int. J. Engg. Techsci. **2**(1), 120–124 (2010)
3. Lee, C.M., Narayanan, S.S.: Toward detecting emotions in spoken dialogs. IEEE Trans. Speech Audio Process. **13**(2), (2005)
4. Ververidis, D., Kotropoulos, C.: Emotional speech recognition: resources, features, and methods, speech communication. Elsevier **48**, 1162–1181 (2006)
5. Fragopanagos, N., Taylor, J.G.: Emotion recognition in human–computer interaction. Neural Networks, Elsevier **18**, 389–405 (2005)
6. Makhoul, J.: Linear prediction: a tutorial review. Proc. IEEE **63**, 561–580 (1975)
7. Ram, R., Palo, H.K., Mohanty, M.N.: Emotion recognition with speech for call centres using LPC and spectral analysis. Int. J. Adv. Comput. Res. **3**(3/11), 189–194 (2013)
8. Quatieri, T.F.: Discrete-Time Speech Signal Processing, 3rd edn. Prentice-Hall, New Jersey (1996)

9. Samal, A., Parida, D., Satpathy, M.R., Mohanty M.N.: On the use of MFCC feature vectors clustering for efficient text dependent speaker recognition. In: Proceedings of International Conference on Frontiers of Intelligent Computing: Theory and Application (FICTA)-2013, Advances in Intelligence System and Computing Series, vol. 247, pp. 305–312. Springer, Switzerland (2014)
10. Palo, H.K., Mohanty, M.N., Chandra M.: Design of neural network model for emotional speech recognition. In: International Conference on Artificial Intelligence and Evolutionary Algorithms in Engineering Systems, April 2014
11. Hermansk, H.: Perceptual linear predictive (PLP) analysis of speech. J. Accoust. Soc. Am. **87**(4), 1739–1752 (1990)
12. Farrell, K.R., Mammone, R.J., Assaleh, K.T.: Speaker networks recognition using neural and conventional classifiers. IEEE Trans. Acoust. Speech Signal Process. **2**(1 part 11), (1994)
13. Javidi, M.M., Roshan, E.F.: Speech emotion recognition by using combinations of C5.0, neural network (NN), and support vector machines (SVM) classification methods. J. Math. Comput. Sci. **6**, 191–200 (2013)

Rectangular Array Languages Generated by a Petri Net

Lalitha D.

Abstract Two different models of Petri net structure to generate rectangular arrays have already been defined. In array token Petri net structure, a transition labeled with catenation rule is enabled to fire only when all the input places of the transition have the same array as token. In Column row catenation Petri net structure, the firing rules differ. A transition labeled with catenation rule is enabled to fire even when different input places of the transition contain different arrays. The firing rule associated with a transition varies in the two models. Comparisons are made between the generative capacity of the two models.

Keywords Array token · Catenation · Inhibitor arcs · Petri net · Picture languages

1 Introduction

Picture languages generated by grammars or recognized by automata have been advocated since the seventies for problems arising in the framework of pattern recognition and image analysis [1–8]. In syntactic approaches to generation of picture patterns, several two-dimensional grammars have been proposed. Array rewriting grammars [6], controlled tabled L-array grammars [5], and pure 2D context-free grammars [8] are some of the picture generating devices. Applications of these models to the generation of "kolam" patterns [9] and in clustering analysis [10] are found in the literature. Oliver Matz's context-free grammars [2] rely on the motion of row and column catenation. The concept of tiling systems (TS) is used as a device of recognizing picture languages [1]. Tile rewriting grammar (TRG) combines Rosenfeld's isometric rewriting rules with the tiling system of Giammarresi [7].

Lalitha D. (✉)
Sathyabama University, Chennai, India
e-mail: lalkrish2007@gmail.com

I.K. Sethi (ed.), *Computational Vision and Robotics*, Advances in Intelligent Systems and Computing 332, DOI 10.1007/978-81-322-2196-8_3

On the other hand, a Petri net is an abstract formal model of information flow [11]. Petri nets have been used for analyzing systems that are concurrent, asynchronous, distributed, parallel, nondeterministic, and/or stochastic. Tokens are used in Petri nets to simulate dynamic and concurrent activities of the system. A string language can be associated with the execution of a Petri net. By defining a labeling function for transitions over an alphabet, the set of all firing sequences, starting from a specific initial marking leading to a finite set of terminal markings, generates a language over the alphabet [11]. The tokens of such a Petri net are just black dots. All the tokens are alike, and the firing rule depends only on the existence of these tokens and the number available in the input places for the transition to fire.

Petri net structure to generate rectangular arrays is found in [12–15]. In [12] column row catenation petri net system has been defined. A transition with several input places having different arrays is associated with a catenation rule as label. The label of the transition decides the order in which the arrays are joined (column-wise or row-wise) provided the condition for catenation is satisfied. In column row catenation petri net system [12], a transition with a catenation rule as label and different arrays in the different input places is enabled to fire. On the contrary in array token Petri nets [13–15], the catenation rule involves an array language. All the input places of the transition with a catenation rule as label should have the same array as token, for the transition to be enabled. The size of the array language to be joined to the array in the input place depends on the size of the array in the input place. Applications of these models to generation of "kolam" patterns [15] and in clustering analysis [13] are found in the literature.

In this paper, we examine the generative capacity of the two different array-generating models. Array token Petri net is able to generate only the regular languages [14]. To control the firing sequence, inhibitor arcs are introduced. The introduction of inhibitor arcs increases the generative capacity. Array token Petri nets with inhibitor arcs generate the context-free and context-sensitive languages [14].

The paper is organized as follows: Sect. 2 and Sect. 3 give the preliminary definitions. Section 4 recalls the concept of array token Petri net structure and column row catenation Petri net system, the language associated with the structure and explains with some examples. Section 5 compares the languages generated by the two different models.

2 Preliminaries

Let Σ be a finite alphabet.

Definition 2.1 A two-dimensional array [6] over Σ is a two-dimensional rectangular array of elements of Σ. The set of all two-dimensional arrays over Σ is denoted by Σ^{**}. A two-dimensional language over Σ is a subset of Σ^{**}.

Two types of catenation operations are defined between two arrays. Let A of size (m, n) and B of size (p, q) be two arrays over an alphabet Σ. The column catenation $A \oplus B$ is defined only when $m = p$, and the row catenation $A \ominus B$ is defined only when $n = q$ $(x)^n$ denotes a horizontal sequence of n "x" and $(x)_n$ denotes a vertical sequence of n "x." $(x)^{n+1} = (x)^n \oplus x$ and $(x)_{n+1} = (x)_n \ominus x$ where $x \varepsilon \Sigma^{**}$.

3 Petri Nets

In this section, the preliminary definitions of Petri Net [11] and notations used are recalled. A Petri net is one of several mathematical models for the description of distributed systems. A Petri net is a directed bipartite graph, in which the bars represent transitions and circles represent places. The directed arcs from places to a transition denote the pre-condition, and the directed arcs from the transition to places denote the post-conditions. Graphically, places in a Petri net may contain a discrete number of marks called tokens. Any distribution of tokens over the places will represent a configuration of the net called a marking. A transition of a Petri net may fire whenever there are sufficient tokens at all the input places. When a transition fires, it consumes these tokens and places tokens at all its output places. When a transition fires, marking of the net changes. Arcs include a "weight" property that can be used to determine the number of tokens consumed or produced when a transition fires. Marking changes according to the firing rules which are given below. In the graph, the weight of an arc is written on the arc.

Definition 3.1 A Petri Net structure is a four-tuple $C = (P, T, I, O)$ where $P = \{P_1, P_2, \ldots, P_n\}$ is a finite set of places, $n \geq 0$, $T = \{t_1, t_2, \ldots, t_m\}$ is a finite set of transitions; $m \geq 0$, $P \cap T = \emptyset$, $I{:}P \times T \to N$ is the input function from places to transitions with weight $w(p, t)$ being a natural number, and O: $T \times P \to N$ is the output function from transitions to places with weight $w(t, p)$ being a natural number.

Note The number of tokens required in the input place, for the transition to be enabled, will depend on the weight of the arc from the place to the transition. The number of tokens put in the output place will depend on the weight of the arc from the transition to the place.

Definition 3.2 A Petri Net marking is an assignment of tokens to the places of a Petri Net. The tokens are used to define the execution of a Petri Net. The number and position of tokens may change during the execution of a Petri Net.

The marking at a given time is shown as a n-tuple, where n is the number of places. The Petri net can be easily represented by a graph. The places are represented by circles, and transitions are represented by rectangular bars. Input and output functions are shown by directed arcs. The weight of an arc is written on the arc in the graph. The weight is assumed to be one if no weight is specified on the arc.

Definition 2.3 An inhibitor arc from a place p_i to a transition t_j has a small circle in the place of an arrow in regular arcs. This means the transition t_j is enabled only if p_i has no tokens. A transition is enabled only if all its regular inputs have tokens and all its inhibitor inputs have zero tokens.

A string language [11] can be associated with the execution of a Petri net. Transitions are labeled with elements of an alphabet. Only the firing sequences that start from a given initial marking and reaching a specific final marking are considered. In this sequence, all the transitions are replaced by their label. This will correspond to a string over the alphabet. Thus, a labeled Petri net generates a string language. Hack [16] and Baker [17] can be referred for Petri net string languages.

4 Petri Net Generating Rectangular Arrays

With labeled Petri net generating string languages as a motivating factor, two models for generating arrays have been introduced. In string generating Petri nets, the tokens are all black dots. A transition is enabled if all the input places have the required number of tokens. In array-generating models, the tokens are arrays over a given alphabet. The firing rules of transition depend not only on the number of the arrays available but also on the size of the array which is residing in the input places of the transition. In Sect. 4.1, the definition of array token Petri net structure [14] is recalled with examples. The definition of column row catenation Petri net [12] is recalled and explained with examples in Sect. 4.2.

4.1 Array Token Petri Net Structure

Array token Petri net structure is defined [14] in such a way that it generates an array language. This structure retains the four components of the traditional Petri net structure $C = (P, T, I, O)$. This Petri net model has places and transitions connected by directed arcs. The marking of the net is just not black dots but arrays over a given alphabet. Rectangular arrays over an alphabet are taken as tokens to be distributed in places. Variation in marking, labels, and firing rules of the transition are listed out below.

Array Language: Array languages are used in the catenation rules. An array language contains an infinite set of arrays. The arrays are having either a fixed number of columns with varying number of rows or having a fixed number of rows with varying number of columns.

Catenation rule as label for transitions: Column catenation rule is in the form $A \ominus B$. The array A denotes the array in the input place of the transition. B is an array language whose number of rows will depend on "m" the number of rows of A. For example, if $B = (x\ x)_m$,then the catenation adds two columns of x after the last column of the array A. $B \ominus A$ would add two columns of x before the first column of A. The number of rows of B is determined by the number of rows of A to enable the catenation.

Row catenation rule is in the form $A \ominus B$. The array A denotes the array in the input place of the transition. B is an array language whose number of columns will depend on "n" the number of columns of A. For example, if $B = \begin{bmatrix} x \\ x \end{bmatrix}^n$, then the catenation $A \ominus B$ adds two rows of x after the last row of the array A. But $B \ominus A$ would add two rows of x before the first row of the array A. The number of columns of B is determined by the number of columns of A to enable the catenation.

Firing rules in array token Petri net

We define the different types of enabled transition in array token Petri net structure. The pre- and post-condition for firing the transition in all the cases are given below:

1. When all the input places of a transition t have the same array as token
   - (i) When the transition t does not have label
     - (a) Each input place should have at least the required number of arrays (dependent on the weight of the arc from the input place to the transition).
     - (b) Firing t consumes arrays from all the input places and moves the array to all its output places
   - (ii) When the transition t has a catenation rule as label
     - (a) Each input place should have at least the required number of arrays (dependent on the weight of the arc from the input place to the transition).
     - (b) The catenation rule is either of the form $A \oplus B$ or $A \ominus B$. The array A denotes the array in the input place. If B is involved in column catenation, then B will be an array language defined with variable row size. If B is involved in row catenation, then B will be an array language defined with variable column size. The variable size of the array B takes a particular value so that the condition for catenation is satisfied.
     - (c) Firing t consumes arrays from all the input places p, and the catenation is carried out in all its output places.
2. When the input places of t have different arrays as token
   - (a) The label of t designates one of its input places.
   - (b) The designated input place has the same array as tokens.
   - (c) The input places have sufficient number of arrays (depends on the weight of the arc from the input place to the transition).
   - (d) Firing t consumes arrays from all the input places and moves the array from the designated input place to all its output places.

Definition 4.1.1 An array token Petri net N is a 8-tuple $N = (P, T, I, O, \Sigma, \sigma, \mu_0, F)$, where $P = \{P_1, \ldots, P_n\}$ is a finite set of places; $T = \{t_1, t_2, \ldots, t_m\}$ is a finite set of transitions; $I : P \times T \rightarrow N$, the input function; $O : P \times T \rightarrow N$, the output function; Σ a finite alphabet; $\sigma : T \rightarrow R\ U\ P$ is a partial function which associates label to certain transitions of the net; R is a set of catenation rules; $\mu_0 : P \rightarrow \Sigma^{**}$ is a partial function which gives an initial marking of arrays in certain places of the net; $F \subseteq P$ is the subset of final places.

Definition 4.1.2 The language generated by an array token Petri net N, $L(N)$, is the set of all rectangular arrays which reach the places belonging to F.

Arrays over a given alphabet are residing in certain places of the net structure. All possible firing sequences are fired in such a way that the arrays move from the initial place to any one of the final places. While the transitions fire either the array just moves from the input place to the output place or it catenates with another array and then moves into the output place. An example is given to explain the concepts.

Definition 4.1.3 An array token Petri net structure with at least one inhibitor arc is defined as array token Petri net with inhibitor arc and is denoted by the N_I.

Definition 4.1.4 The family of array languages generated by array token Petri net is denoted by $L(N)$, and the family of array languages generated by array token Petri net with inhibitor arc is denoted by $L(N_1)$.

Example 4.1.1 The array token labeled Petri net $N_I(1) = (P, T, I, O, \Sigma, \sigma, \mu_0, F)$, where $P = \{P_1, P_2, P_3, P_4, P_5, P_6, P_7\}$, $T = \{t_1, t_2, t_3, t_4, t_5, t_6\}$. The input and output arrows are seen from the figure. The labels are shown in the figure. $\Sigma = \{a\}$. $\mu_0(p_1) = \mu_0(p_2) = \mu_0(p_7) = $ s is the initial marking. $F = \{P_7\}$ is the final place (Fig. 1).

The arrays used are $S = \begin{matrix} a & a \\ a & a \end{matrix}$, $B_1 = \begin{pmatrix} a & a \end{pmatrix}_m$, $B_2 = \begin{pmatrix} a \\ a \end{pmatrix}^n$. To start with t_1 is enabled. Firing t_1 pushes the 2×2 array S in P_2 to P_3. Firing t_2 adds two columns and firing t_3 adds two rows.

Firing the sequence $t_1t_2t_3$ generates the array

$$\begin{matrix} a & a & a & a \\ a & a & a & a \\ a & a & a & a \\ a & a & a & a \end{matrix}$$

in P_2 and P_5. At this stage, there is no array in P_1, and so both t_4 and t_6 are enabled. Firing t_6 pushes the 4×4 array into P_7 and P_2. Firing t_4 enables t_5. Since

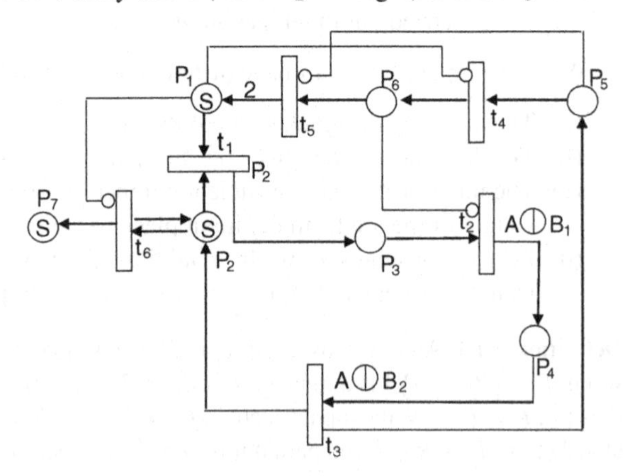

Fig. 1 Petri net to generate square array of side 2^n

weight of the arc from t_5 to P_1 is two, firing t_5 pushes two 4×4 arrays into P_1. Now, the sequence $t_1t_2t_3$ has to be fired two times for both t_4 and t_6 to be enabled. Then, the array in P_2 can be pushed into P_7. Thus, $t_1t_2t_3t_7t_4t_5\ (t_1t_2t_3)^2\ t_6$ generates the array of size 8×8. The language generated is squares of side 2^n, $n \geq 1$.

4.2 Column Row Catenation Petri Net Structure

Column row catenation Petri net system [12] also generates an array language. This structure retains the eight components $N = (P, T, I, O, \Sigma, \sigma, \mu_0, F)$ of array token Petri net structure. When the input places have different arrays, the transition only shifts the array that is specified by the label of the transition in the previous model. But in column row catenation Petri net, the transition consumes the different arrays and catenates them (condition for catenation has to be satisfied). This variation in labels and firing rules of the new model are listed out below.

In this model, a variation is made in the catenation rule when the transition consumes different arrays from the input places. The label of the transition could be a catenation rule. Let A and B be two different arrays with same number of columns in the two input places P_1 and P_2 of a transition t. Then, the catenation rule $P_1 \ominus P_2$ can be given as a label of the transition t. The transition t, on firing, will join A and B row-wise in the same order. Similarly, if C and D are two different arrays with same number of rows in the two input places P_1 and P_2 of a transition t. Then, the catenation rule $P_1 \oplus P_2$ can be given as a label of the transition t. The transition t, on firing, will join C and D column-wise in the same order. The number of input places is not restricted. The catenation rule specifies the order in which the arrays are joined (row-wise or column-wise). $P_1 \ominus P_2 \ominus \ldots \ominus P_k$ joins the arrays in $P_1, P_2, \ldots, P_k$ row-wise in that order.

Firing rules in Column row catenation Petri net structure

Both the pre- and post-condition of the firing rules enlisted 1(i) and 1(ii) in array token Petri net structure holds in this model. The firing rules are the same when all the input places of a transition have the same array as token. When the input places have different arrays as tokens, then the firing rules differ.

Input places having different arrays as tokens can be classified as follows:

(i) Within one input place different arrays can reside.
(ii) Different input places have different arrays but within one input place only copies of the same array is found.

Firing rules when the input places have different arrays as token.

1. When at least one input place of t has different arrays as token and if label of t is one of the input places
   (a) The input place, designated by the label, should have the same array as tokens.

(b) The input places have sufficient number of arrays (depends on the weight of the arc from the input place to the transition).
(c) Firing t consumes arrays from the input places and moves the array from the designated input place to all its output places.

2. When different input places of t have different arrays, but within one input place only copies of the same array is found and if label of t is a catenation rule

(a) The input places have sufficient number of arrays (depends on the weight of the arc from the input place to the transition).
(b) The catenation rule is either of the form $P_1 \ominus P_2 \ominus \ldots \ominus P_k$ or $P_1 \oplus P_2 \oplus \ldots \oplus P_k$.
(c) The arrays should satisfy the condition for catenation.
(d) Firing t consumes arrays from all the input places and joins the arrays in the order stated and puts in all its output places.

Definition 4.2.1 A column row catenation Petri net structure N_{CR} is an 8-tuple $N_{CR} = (P, T, I, O, \Sigma, \sigma, \mu_0, F)$, where $P = \{p_1,\ldots, p_n\}$ is a finite set of places; $T = \{t_1, t_2,\ldots, t_m\}$ is a finite set of transitions; $I : P \times T \rightarrow N$, the input function; $O : P \times T \rightarrow N$, is the output function; Σ a finite alphabet; $\sigma : T \rightarrow R\ U\ P$ is a partial function which associates label to certain transitions of the net, R is a set of catenation rules; $\mu_0 : P \rightarrow \Sigma^{**}$ is a partial function which gives an initial marking of arrays in certain places of the net; $F \subseteq P$ is the subset of final places.

Definition 4.2.2 The language generated by a column row catenation Petri net structure N_{CR}, $L(N_{CR})$, is the set of all rectangular arrays which reach the places belonging to F.

Definition 4.2.3 The family of such array languages generated by column row catenation Petri net structure is denoted by $L(N_{CR})$

Example 4.2.1 The column row catenation Petri net structure is the 8-tuple $N_{CR}(1) = (P, T, I, O, \Sigma, \sigma, \mu_0, F)$ with $P = \{P_1, P_2, P_3, P_4, P_5, P_6, P_7, P_8, P_9\}$, $T = \{t_1, t_2, t_3, t_4, t_5, t_6, t_7, t_8, t_9\}$. The input and output function are seen from the graph given in Fig. 2. $\Sigma = \{*, \bullet\}$. The labels are seen in the graph. $\mu_0(P_1) = \mu_0(P_4) = S$ is the initial marking. $F = \{P_9\}$ is the final place. The arrays involved are

$$s = \begin{matrix} \bullet \\ * \\ \bullet \end{matrix}\ , B_1 = (*)_m,\ B_2 = (\bullet)^n$$

Firing the sequence of transition $t_1t_2t_3$, say three times puts the array given in Fig. 3a in P_7 and firing the sequence $t_5t_5t_6$ puts the array given in Fig. 3b in P_8, and firing the transition generates the diamond of size 9×9 given Fig. 3c in P_9, t_9 the final place. The language generated by $N_{CR}(1)$ is the set of all diamonds of odd size length.

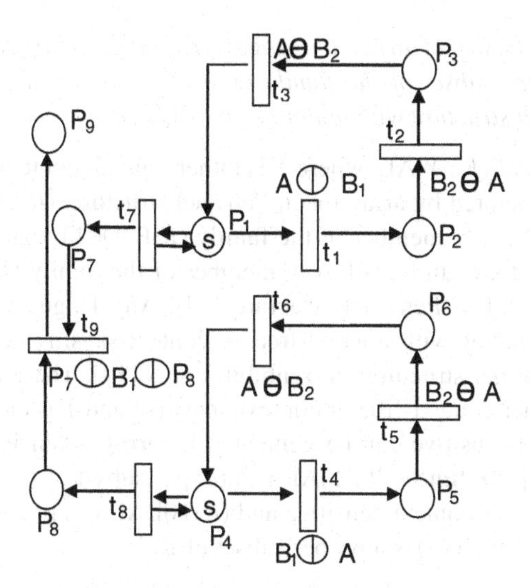

Fig. 2 Petri net to generate diamonde of odd side length

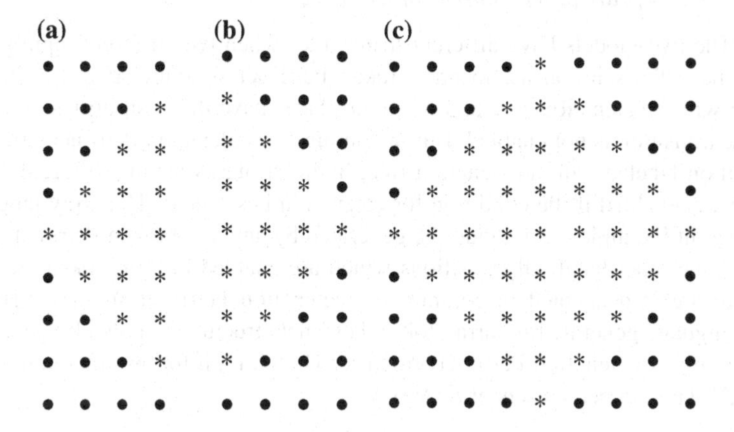

Fig. 3 The arrays reaching the places P_7, P_8, P_9 respectively

5 Generative Capacity of the Models

In this section, comparison of the generative capacity of the two models is done with the other array language generating models. For the definition of array rewriting grammar refer to [6], extended controlled table L-array grammar refer to [5], and pure 2D context-free grammar refer to [8].

Theorem 5.1 *The family of array languages generated by array token Petri net structure is a proper subset of the family of such array languages generated by array token Petri net structure with inhibitor arc.* $\mathcal{L}(\mathcal{N})$ *is a proper subset of* $\mathcal{L}(\mathcal{N}_{\mathcal{I}})$.

Proof The families of $(R : Y)$AL, where Y is either regular, context-free or context-sensitive can be generated by array token Petri net structure. $(R : Y)$AL is a proper subset of $\mathcal{L}(\mathcal{N})$ [14]. Any member of the family (R)P2DCFL can be generated by array token Petri net structure [14]. Any member of the family (R)T0LAL can be generated by array token Petri net structure [14]. Any language generated by a Table 0L-array grammar with context-free or context-sensitive can be generated by array token Petri net structure with inhibitor arcs [14]. The families of $(X : Y)$ AL, where X is either context-free or context-sensitive and Y is either regular, context-free, or context-sensitive can be generated by array token Petri net structure with inhibitor arcs [14]. Since $(R : Y)$AL is a proper subset of $(X : Y)$AL, where X is either context-free or context-sensitive and (R)T0LAL is a proper subset of (CF) T0LAL, it follows that $\mathcal{L}(\mathcal{N})$ is a proper subset of $\mathcal{L}(\mathcal{N}_{\mathcal{I}})$.

Theorem 5.2 *The family of array languages generated by array token Petri net structure (with or without inhibitor arc) is a proper subset of the family of such array languages generated by column row catenation Petri net structure. Both* $\mathcal{L}(\mathcal{N})$ *and* $\mathcal{L}(\mathcal{N}_{\mathcal{I}})$ *are proper subsets of* $\mathcal{L}(\mathcal{N}_{\mathcal{CR}})$.

Proof The two models have different firing rules when two different input places do not have the same array. In array token Petri net structure, if a transition is labeled with a catenation rule and its input places have different arrays as token, then the transition is not enabled. But in column row catenation Petri net structure, a transition labeled with a catenation rule, its input places having different arrays as token, is enabled if the condition for catenation is satisfied. The array language generated in Example 4.2.1 cannot be generated by any array token Petri net structure. On the other hand, all transitions which are enabled in array token Petri net structure are also enabled in column row catenation Petri net structure. Hence, every language generated by array token Petri net structure can also be generated by column row catenation Petri net structure. Therefore, it follows that both $\mathcal{L}(\mathcal{N})$ and $\mathcal{L}(\mathcal{N}_{\mathcal{I}})$ are proper subsets of $\mathcal{L}(\mathcal{N}_{\mathcal{CR}})$.

6 Conclusion

Examples have been given to analyze the differences in the firing rules and the generative capacity of the two models. The generative capacity of the array token Petri net structure with inhibitor arc is more than the generative capacity of array token Petri net structure. It has been proved that the column row catenation structure has more generative capacity than the array token Petri net structure.

References

1. Giammarresi, D., Restivo, A.: Two-dimensional languages. In: Rozenberg, G., Salomaa, A. (eds.) Handbook of Formal Languages, vol. 3, pp. 215–267. Springer, Berlin (1997)
2. Matz, O.: Regular expressions and context-free grammars for picture language. In: Proceedings of 14th Annual Symposium on Theoretical Aspects of Computer Science, LNCS, vol. 1200, pp. 283–294 (1997)
3. Rosenfeld, A., Siromoney, R.: Picture languages—a survey. Lang. Des. **1**(3), 229–245 (1993)
4. Siromoney, R.: Advances in array languages. In: Proceedings of the 3rd International Workshop on Graph Grammars and Their Application to Computer Science. LNCS, vol. 291, pp. 549–563 (1987)
5. Siromoney, R., Siromoney, G.: Extended controlled table L-arrays. Inf. Control **35**, 119–138 (1977)
6. Siromoney, G., Siromoney, R., Kamala, K.: Picture languages with array rewriting rules. Inf. Control **22**, 447–470 (1973)
7. Reghizi, S.C., Pradella, M: Tile rewriting grammars and picture languages. TCS **340**, 257–272 (2005)
8. Subramanian, K.G., Rosihan M. Ali, Geethalakshmi, M., Nagar, A.K.: Pure 2D picture grammars and languages. Discrete Appl. Math. **157**(16), 3401–3411 (2009)
9. Siromoney, G., Siromoney, R., Kamala, K.: Array grammars and kolam. Comput. Graph. Image Process. **3**(1), 63–82 (1974)
10. Wang, P.S.P.: An application of array grammars to clustering analysis for syntactic patterns. Pattern Recogn. **17**(4), 441–451 (1984)
11. Peterson, J.L.: Petri Net Theory and Modeling of Systems. Prentice Hall Inc, Englewood Cliffs (1981)
12. Lalitha, D., Rangarajan, K.: Column and row catenation petri net systems. In: Proceedings of the Fifth IEEE International Conference on Bio-inspired Computing: Theories and Applications, pp. 1382–1387 (2010)
13. Lalitha, D., Rangarajan, K.: An application of array token Petri nets to clustering analysis for syntactic patterns. Int. J. Comput. Appl. **42**(16), 21–25 (2012)
14. Lalitha, D., Rangarajan, K., Thomas, D.G.: Rectangular arrays and Petri nets. Comb. Image Anal. LNCS **7655**, 166–180 (2012)
15. Lalitha, D., Rangarajan, K.: Petri nets generating kolam patterns. Indian J. Comput. Sci. Eng. **3**(1), 68–74 (2012)
16. Hack, M.: Petri net languages. Computation structures group memo 124, Project MAC, MIT (1975)
17. Baker, H.G.: Petri net languages. Computation structures group memo 68, Project MAC, MIT, Cambridge, Massachusetts (1972)

Medical Image Fusion Using Local Energy in Nonsubsampled Contourlet Transform Domain

Richa Srivastava and Ashish Khare

Abstract Image fusion is an emerging area of image processing. It integrates complementary information of different source images into a single fused image. In the proposed work, we have used nonsubsampled contourlet transform for fusion of images which is a shift-invariant version of contourlet transform. Along with this property, it has many advantages like removal of pseudo-Gibbs phenomenon, better frequency selectivity, improved temporal stability, and consistency. These properties make it suitable for fusion application. For fusing images, we have used local energy-based fusion rule. This rule depends on the current as well as the neighboring coefficients. Hence, it performs better than single coefficient-based fusion rules. The performance of the proposed method is compared visually and quantitatively with contourlet transform, curvelet transform, dual-tree complex wavelet transform, and Daubechies complex wavelet transform-based fusion methods. To evaluate the methods quantitatively, we have used mutual information, edge strength, and fusion factor quality measurements. The experimental results show that the proposed method performs better and is more effective than other methods.

Keywords Image fusion · Nonsubsampled contourlet transform · Shift invariance · Local energy-based fusion

R. Srivastava (✉) · A. Khare
Department of Electronics and Communication, University of Allahabad, Allahabad, India
e-mail: gaur.richa@gmail.com

A. Khare
e-mail: ashishkhare@hotmail.com

I.K. Sethi (ed.), *Computational Vision and Robotics*, Advances in Intelligent Systems and Computing 332, DOI 10.1007/978-81-322-2196-8_4

1 Introduction

Image fusion [1, 2] is a technique that integrates complementary information, scattered in different images into a single composite image. The fused image contains more information than any of its source images with reduced noise and artifacts. Hence, the results of the fusion are more useful for human perception and further image processing tasks. The idea behind this technique is the sensors of limited capacity that cannot capture the complete information of a scene. For example, in medical imaging [1], the computed tomography (CT) image is suitable for bone injuries, lungs, and chest problems, whereas the magnetic resonance imaging (MRI) is used for examining soft tissues like spinal cord injuries, brain tumors, etc. Various fusion methods are developed by the researchers. These methods are classified into pixel-level [3], feature-level [4], and decision-level fusion [5]. Pixel-level fusion is again divided into spatial- and transform-domain fusion. Spatial-domain fusion methods (averaging, weighted averaging, principal component analysis) are simple but sometimes give poor fusion results. Transform-domain fusion methods avoid the problems of spatial-domain methods and give better results. Different transforms are developed by the researchers such as Laplacian pyramid, gradient pyramid, wavelet transform [6], and contourlet transform [7]. Wavelet transform is very popular among the researchers, but it has limited number of directions, due to which it is not efficient in handling two-dimensional singularities. Contourlet transform, proposed by Do and Vetterli [7], provides flexible number of directions but is shift variant. Shift variance causes many problems like pseudo-Gibbs phenomenon near singularity. To tackle this problem, Chunha et al. [8] gave a multiscale, multidirectional, and shift-invariant transform known as nonsubsampled contourlet transform (NSCT). It has better frequency selectivity and improved temporal stability, and it removes the problem of pseudo-Gibbs phenomenon that causes smoothness near singularities. All these properties make it a suitable transform for image fusion.

In this work, we have proposed a local energy-based image fusion in NSCT domain. The source images are decomposed in transform domain by applying NSCT. Then, we have calculated local energy of each coefficient and then compared the local energy of corresponding coefficients of source images. The coefficient with highest local energy is selected. After obtaining the coefficient set for the fused image, the inverse NSCT is applied to get the final fused image.

The rest of the paper is organized as follows: Sect. 2 gives the basic concept of NSCT. Section 3 describes the proposed method in detail. Experimental Results and discussion are discussed in Sect. 4. Finally, the paper is concluded in Sect. 5.

2 Nonsubsampled Contourlet Transform (NSCT)

NSCT [9, 10] is a multiscale, multidirectional, and geometrical transform. Unlike contourlet transform, it is shift invariant. In contourlet, the shift variance is caused by the downsampling in Laplacian pyramid filter banks (LPFB) as well as in directional filter banks (DFB) [11]. NSCT achieves shift invariance by removing the downsampling and upsampling from the LPFB and DFB. The construction of NSCT is the combination of nonsubsampled pyramidal filter banks (NSPFB) and nonsubsampled directional filter banks (NSDFB). The NSPFB provides the multiscale property, whereas the NSDFB gives multidirectional property to NSCT. Its implementation is similar to that of nonsubsampled wavelet transform obtained by a`trous algorithm. Here, the filter of the next decomposition level is obtained by upsampling the filter of previous stage. Hence, multiscale property is achieved easily without extra filter design algorithm.

The DFB is proposed by Bamberger and Smith [11]. It is obtained by incorporating the critically sampled two-channel fan filter banks and resampling operator. The DFB has tree-like structure which splits the frequency plane in directional wedges. But it is not shift invariant. To make it shift invariant, the downsampler and upsampler are removed. This results in tree-shaped two-channel nonsubsampled filter banks.

The NSCT is constructed by combining the NSPFB and NSDFB. It is invertible and satisfies the anisotropic scaling law. Also, the removal of downsampling makes the filter designing simpler.

3 The Proposed Method

In this paper, we have proposed local energy-based image fusion in NSCT domain. In local energy-based fusion, the decision of coefficient selection is based not only on the coefficient itself but also on its neighboring coefficients. Hence, local energy-based fusion gives more efficient results than single coefficient-based fusion rules such as maximum selection or average fusion rules. The local energy is calculated by taking absolute sum of the neighboring coefficients of the current coefficient. Then, we compare the local energy of the coefficients of the source images and select the coefficient that has higher value of local energy.

In the proposed method, we have performed fusion in NSCT domain. It has better directional selectivity and is shift invariant. Also, it avoids the pseudo-Gibbs phenomenon, which is a big problem in wavelet transform- and contourlet transform-based fusion methods. The proposed algorithm can be summarized as below:

Step 1 The source images are decomposed into coefficient sets using the NSCT.

$$Im_1(i,j) \overset{\text{NSCT}}{\rightarrow} Cf_1(i,j) \quad Im_2(i,j) \overset{\text{NSCT}}{\rightarrow} Cf_2(i,j)$$

Step 2 Local energy of each coefficient is calculated in 3×3 window using the following formula:

$$e_1(i,j) = \sum_{i-1}^{i+1} \sum_{j-1}^{j+1} Cf_1(i,j) \quad e_2(i,j) = \sum_{i-1}^{i+1} \sum_{j-1}^{j+1} Cf_2(i,j)$$

Step 3 The coefficient with higher local energy is selected from the two.

$$Cf(i,j) = \begin{cases} Cf_1(i,j) & \text{if } |e_1(i,j)| \geq |e_2(i,j)| \\ Cf_2(i,j) & \text{otherwise} \end{cases}$$

Step 4 Reconstruct the final fused image by applying inverse NSCT on the above coefficients.

$$Cf(i,j) \overset{\text{Inverse NSCT}}{\rightarrow} F(i,j)$$

4 Experimental Results and Discussion

In this section, we have shown the results of the proposed method and its comparison with other methods. For experiment, we have taken two different multimodal medical images (Figs. 1a, b and 2a, b) of size 256 × 256. Each set contains one CT image and one MRI. To show the effectiveness of the proposed method, it is compared with curvelet transform [12], contourlet transform, and dual-tree complex wavelet transform [13] and Duabechies complex wavelet transform-based [14] methods with three different fusion rules (absolute maximum, local energy, and edge preserving) [10, 15, 16]. The visual results of the proposed and other methods are shown in Figs. 1 and 2. From Figs. 1 and 2, we see that the proposed

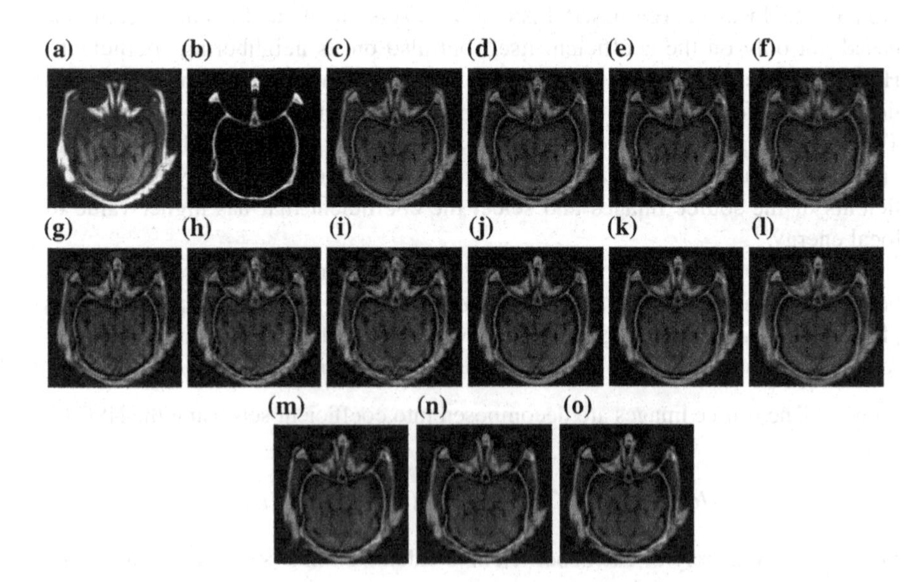

Fig. 1 Fusion results for first set of multifocus images. **a** Image 1. **b** Image 2. **c** The Proposed Method. **d** Cvt (Max). **e** Cvt (Energy). **f** Cvt (Edge). **g** CT (Max). **h** CT (Energy). **i** CT (Edge). **j** DTCWT (Max). **k** DTCWT (Energy). **l** DTCWT (Edge). **m** DCxWT (Max). **n** DCxWT (Energy). **o** DCxWT (Edge)

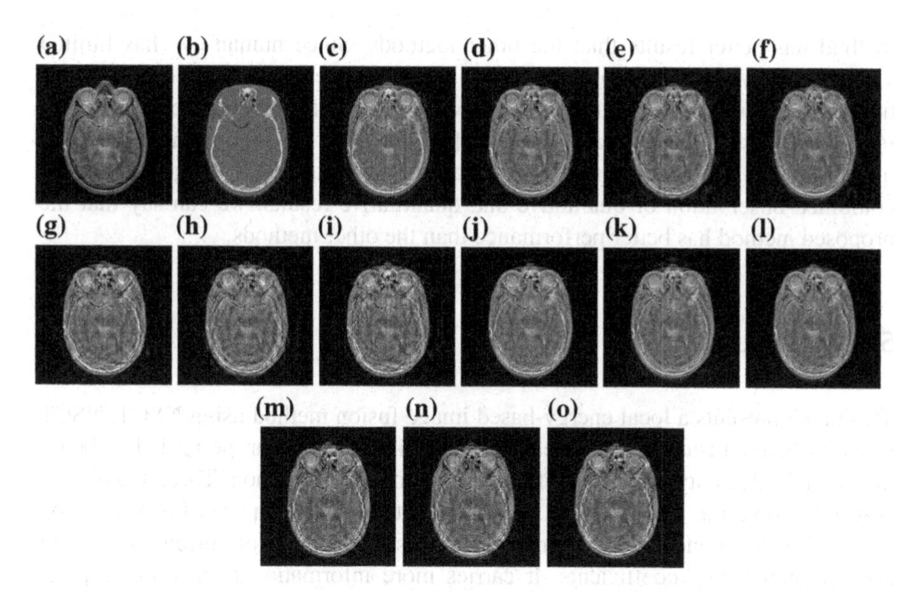

Fig. 2 Fusion results for second set of multifocus images. **a** Image 1. **b** Image 2. **c** The Proposed Method. **d** Cvt (Max). **e** Cvt (Energy). **f** Cvt (Edge). **g** CT (Max). **h** CT (Energy). **i** CT (Edge). **j** DTCWT (Max). **k** DTCWT (Energy). **l** DTCWT (Edge). **m** DCxWT (Max). **n** DCxWT (Energy). **o** DCxWT (Edge)

Table 1 Fusion results for both set of medical images

	First set of medical image			Second set of medical image		
Methods	Mutual information	Q_{AB}^{F}	Fusion factor	Mutual information	Q_{AB}^{F}	Fusion factor
The Proposed Method	4.077303	0.756979	3.707573	3.895457	0.560030	3.958168
Cvt (Max)	1.969366	0.612996	1.437993	2.897573	0.525046	2.891028
Cvt (Energy)	1.815109	0.578114	1.335082	2.852393	0.500984	2.854308
Cvt (Edge)	1.886123	0.541534	1.393054	2.907971	0.469633	2.919603
CT (Max)	1.741266	0.522747	1.279253	2.824510	0.430833	2.739439
CT (Energy)	1.656684	0.483646	1.224120	2.795204	0.415443	2.731459
CT (Edge)	1.632673	0.473959	1.196554	2.766341	0.374957	2.696326
DTCWT (Max)	2.632675	0.695092	1.903239	3.282068	0.567474	3.368549
DTCWT (Energy)	2.660842	0.676518	1.918290	3.249007	0.556403	3.510635
DTCWT (Edge)	2.058280	0.621017	1.494364	3.214950	0.554437	3.432166
DCxWT (Max)	1.854965	0.616381	1.205011	3.046493	0.508618	3.033609
DCxWT (Energy)	1.951738	0.597656	1.326218	3.014638	0.483526	3.019793
DCxWT (Edge)	1.854965	0.616381	1.205011	3.046493	0.508618	3.033609

method has better results than the other methods. Since human eye has limited vision capacity, it cannot observe the minute difference in the images. Hence, we have also evaluated the results quantitatively. Mutual information, edge strength (Q^F_{AB}), and fusion factor quality metrics [17] are used for comparing the results quantitatively. Results are presented in Table 1 for both set of images. From the combined observation of qualitative and quantitative results, we can say that the proposed method has better performance than the other methods.

5 Conclusions

This paper presents a local energy-based image fusion method using NSCT. NSCT provides better fusion results because of its shift-invariant property. It has better directional selectivity and avoids the pseudo-Gibbs phenomenon. These features of NSCT improve the results of the proposed method. The proposed fusion method is based on local energy. Local energy contains information of current coefficient and its neighboring coefficients. It carries more information than a single pixel has. Hence, it is more reliable and efficient than single pixel-based fusion rules. To prove the effectiveness of the proposed method, it is compared with curvelet transform, contourlet transform, dual-tree complex wavelet transform, and Daubechies complex wavelet transform-based fusion methods. Three standard quantitative measurements—mutual information, edge strength, and fusion factor—are used for evaluating the results quantitatively. Both the visual and quantitative results show that the proposed method has better performance than the curvelet, contourlet, dual-tree complex, and Daubechies transform-based fusion methods.

References

1. James, A.P., Dasarathy, B.V.: Medical image fusion: a survey of state of the art. Inf. Fusion **19**, 4–19 (2014)
2. Khalegi, B., Khamis, A., Karray, F.O., Razavi, S.N.: Multisensor data fusion: a review of the state-of-the-art. Inf. Fusion **14**(1), 28–44 (2013)
3. Yang, B., Li, S.: Pixel level image fusion with simultaneous orthogonal matching pursuit. Inf. Fusion **13**(1), 10–19 (2012)
4. Kannan, K., Perumal, S.A., Arulmozhi, K.: The review of feature level fusion of multi-focused images using wavelets. Recent Pat. Signal Process **2**, 28–38 (2010)
5. Prabhakar, S., Jain, A.K.: Decision-level fusion in fingerprint verification. Pattern Recogn. **35**(4), 861–874 (2002)
6. Pajares, G., Cruz, J.: A wavelet based image fusion tutorial. Pattern Recogn. **37**(9), 1855–1872 (2004)
7. Do, M.N., Vetterli, M.: The contourlet transform: an efficient directional multiresolution image representation. IEEE Trans. Image Process. **14**(12), 2091–2106 (2005)
8. Chunha, L., Zhou, J., Do, M.N.: The nonsubsampled contourlet transform: theory, design and applications. IEEE Trans. Image Process. **15**(10), 3089–3310 (2006)
9. Fu, L., Yifan, L., Xin, L.: Image fusion based on nonsubsampled contourlet transform and pulse coupled neural networks. In: 4th International Conference on Intelligent Computation Technology and Automation, pp. 572–575. IEEE press, Guangdong (2011)

10. Srivastava, R., Singh, R., Khare A.: Image fusion based on nonsubsampled contourlet transform. In International Conference on Informatics, Electronics and Vision, pp. 263–266. IEEE press, Dhaka
11. Bamberger, R.H., Smith, M.J.T.: A filter bank for the directional decomposition of images: theory and design. IEEE Trans. Signal Process. **40**(4), 882–893 (1992)
12. Nehcini, F., Garzelli, A., Baronti, S., Alparone, L.: Remote sensing image fusion using the curvelet transform. Inf. Fusion **8**(2), 143–156 (2007)
13. Hui, X., Yihui, Y., Benkang, C., Yiyong, H.: Image fusion based on complex wavelets and region segmentation. In: International conference on Computer Application and System Modeling (ICCASM), pp. 135–138. IEEE press, Taiyuan (2010)
14. Singh, R., Khare, A.: Fusion of multimodal medical images using Daubechies complex wavelet transform—a multiresolution approach. Inf. Fusion **19**, 49–60 (2014)
15. Khare, A., Srivastava, R., Singh, R.: Edge preserving image fusion based on contourlet transform. In: 5th International Conference on Image and Signal Processing (ICISP), pp. 93–102. Springer, Morocco (2012)
16. Lu, H., Li, Y., Kitazono, Y., Zang, L.: Local energy based multi-focus image fusion on curvelet transform. In: International Symposium on Communication and Information Technologies (ISCIT), pp. 1154–1157. IEEE press, Tokyo (2010)
17. Xydeas, C.S., Petrovic, V.: Objective image fusion performance measure. Electron. Lett. **36**(4), 308–309 (2000)

Machine Learning and Its Application in Software Fault Prediction with Similarity Measures

Ekbal Rashid, Srikanta Patnaik and Arshad Usmani

Abstract Nowadays, the challenge is how to exactly understand and apply various techniques to discover fault from the software module. Machine learning is the process of automatically discovering useful information in knowledgebase. It also provides capabilities to predict the outcome of future solutions. Case-based reasoning is a tool or method to predict error level with respect to LOC and development time in software module. This paper presents some new ideas about process and product metrics to improve software quality prediction. At the outset, it deals with the possibilities of using lines of code and development time from any language may be compared and be used as a uniform metric. The system predicts the error level with respect to LOC and development time, and both are responsible for checking the developer's ability or efficiency of the developer. Prediction is based on the analogy. We have used different similarity measures to find the best method that increases the correctness. The present work is also credited through introduction of some new terms such as coefficient of efficiency, i.e., developer's ability and normalizer. In order to obtain the result, we have used indigenous tool.

Keywords Software fault prediction · Similarity function · LOC · Development time · Normalizer

E. Rashid (✉) · A. Usmani
Cambridge Institute of Technology, Ranchi, India
e-mail: ekbalrashid2004@yahoo.com

A. Usmani
e-mail: ausmani@yahoo.co.in

S. Patnaik
SOA University, Bhubaneswar, Orissa, India
e-mail: patnaik_srikanta@yahoo.co.in

I.K. Sethi (ed.), *Computational Vision and Robotics*, Advances in Intelligent Systems and Computing 332, DOI 10.1007/978-81-322-2196-8_5

1 Introduction

At the present time, various machine learning techniques available for software fault prediction. Artificial intelligence (AI)-based techniques such as case-based reasoning (CBR), genetic algorithms, and neural networks have been used for fault prediction. But the biggest challenge is how to really apply these techniques in fault prediction, and which technique is more effective with respect to time. CBR is a most popular machine learning technique [1]. Currently, the major application areas of CBR are in the field of health sciences and multiagent systems as well as in Web-based planning. Research on CBR in these areas is growing, but most of the systems are still prototypes and not available in the market as commercial products. However, many of the systems are intended to be commercialized [2]. The novel idea of CBR is that 'similar problems have similar solutions.' There have been several models for CBR that attempt to offer better understanding of CBR [3]. Some authors shown that CBR is a methodology and not a technology [4] and it is a methodology for solving the new problems [5]. Rashid et al. [6] emphasized on the importance of machine learning and software quality prediction as an expert system. On the other hand, software quality is a set of characteristics that can be measured in all phases of software development. Software quality can be broadly divided into two categories, namely quality of product and quality of process.

(a) Quality of product: Quality of product provides a measure of a software product during the process of development which can be in terms of lines of code or source code.
(b) Quality of process: Quality of process is a measure of the software development process such as effort, development and time

Quality software is reasonable bug-free, delivered on time and within budget, meets requirements, and is maintainable.

The rest of the paper is structured as follows: Sect. 2 gives methodology, Sect. 3 describes the model of CBR. In Sect. 4, we present similarity functions. In Sect. 5, we present error level calculation, and Sect. 6 presents Results and Discussion.

1.1 LOC

The lines of code continue to be the determining metric in most of the quality and cost measurement. Now the method in which the LOC count has to be taken is also not standardized as yet. However, if we for the sake of simplicity assume that a standardization has been achieved, then on the basis of those set standards, after taking the LOC count, we can normalize the same. This standard LOC count after being normalized serves as the primary metric [7].

Table 1 The comparison of developer's ability in module 1 and module 2

Module	Man-months	LOC	Experience (in years)
Module 1	4	40,000	5
Module 2	6	41,000	3

1.2 Development Time

The development time can also be a confusing factor. With experienced personnel, the development time is sure to be on the lower side. However, if the staff is not equipped enough, we are going to have a prolonged development time with respect to any particular work. So this development time also needs to be brought on a particular level so that the ambiguity vanishes totally. For this purpose, a parameter called the coefficient of efficiency may be introduced. It is the duty of the organization to rate different developers with the different coefficients, and this rating will establish the relative efficiency of one over the other. The developer with the lowest efficiency can be rated 1. Subsequently, the other developers may be rated in increasing order of merit. The actual clocked time required to complete a chunk of work or a module may be multiplied with the reciprocal of this coefficient of efficiency to get the actual development time for the particular piece of work [2]. It may be expressed as below:

$$\text{Normalized development time} = \text{clocked development time} \times \frac{1}{\text{coefficient of efficiency}(\alpha)}$$

where α is the constant of proportionality and may be called the coefficient of the developer's ability or efficiency. Thus, we can define coefficient of developer's ability (CDA) as follows:

CDA = LOC/man-month

where CDA = coefficient of developer's ability

LOC = lines of code

Looking at Table 1, it is evident that developer's ability engaged in module 1 is more than the developer's ability engaged in module 2. Therefore, it infers that experienced person takes less time in comparison with inexperience person.

1.3 The Normalizer

Let the parameter or coefficient that shows the equivalence of lines of code written in any language with the lines of code written in assembly language be called normalizer. It is not very easy to determine the normalizer. Leave aside determining the normalizer; it is also very difficult to make an idea of how to determine the normalizer. One of the better ways possible is by assigning some coefficients to the physical lines of code counted in different languages. If we just take up the simple idea that as the languages evolved, they actually targeted to do away with excessive coding. The assembly code executes one instruction for every single line of code, and so it can be taken as the basis of all calculation. Its coefficient can be set to 1. Next as the higher languages developed, we can assign coefficients in the increasing order. A separate set of coefficients can be set for structural and object-oriented languages.

2 Methodology

It is very difficult to get industrial data for research in the field of software engineering domain. Therefore, in this research, the environment of our study is the college campus and our target group is the B. Tech students of computer science and engineering. Students are guided with their faculty, supported by the technical staffs, resources such as computers and software. In this section, I was personally involved for collecting the data from the B. Tech computer science students.

The values for lines of code, number of functions, level of difficulty, experience, number of variables, and development time were collected from programs developed by B. Tech students of the college campus over a period of six months. Measurement of software quality (fault) prediction has been done on CBR basis.

3 Model of CBR

Depending upon the problems provided, the best matches are found from the knowledgebase. See Figs. 1, 2, and 3, respectively.

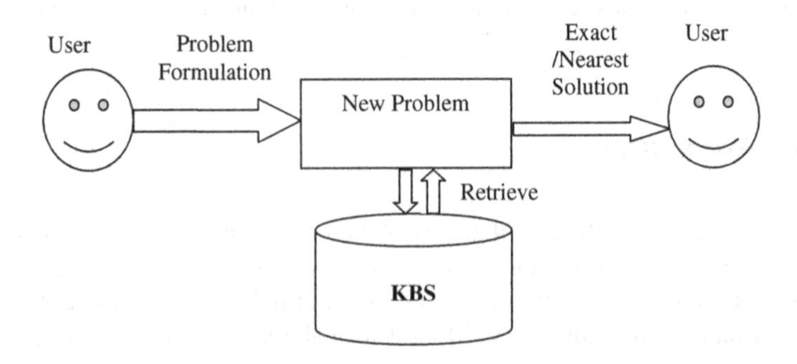

Fig. 1 Context-level diagram

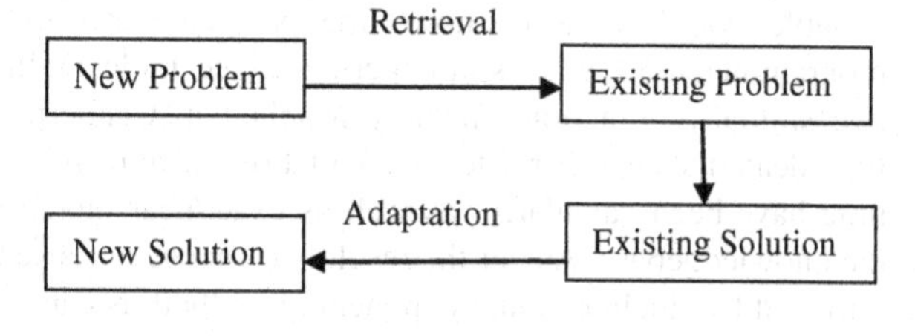

Fig. 2 Reuse principle

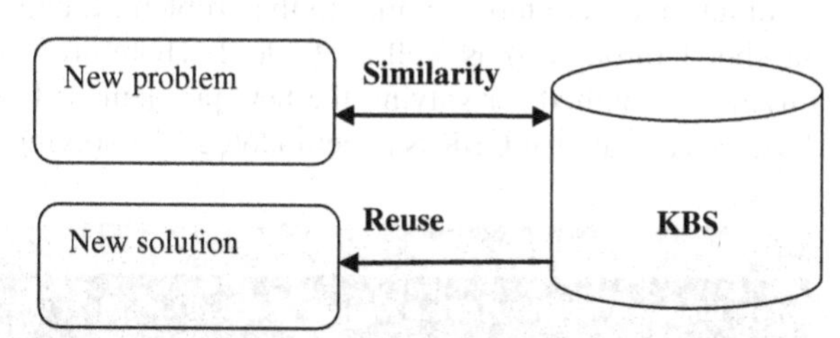

Fig. 3 Conceptual CBR procedure

4 Similarity Functions Used

Similarity has been used for retrieval purposes. The difference between the input case and the retrieved case can be found out about the concept of the Euclidian distance, Manhattan distance, Clark distance, Canberra distance, and exponential distance. These measures are used to calculate the distance of the new record set or case from each record set stored in the knowledgebase. The matching case(s) are those that have the minimum distance from the new record set [8].

5 Error Level Calculation

We can proceed to calculate the error level of the module or the particular section of the software as below:
With respect to development time:

$$\text{ERROR LEVEL (EL)} = \frac{\text{number of errors}}{\text{development time}} \tag{1}$$

With respect to LOC:

$$\text{ERROR LEVEL (EL)} = \frac{\text{number of errors}}{\text{LOC}} \tag{2}$$

6 Results and Discussion

In this research work, we have used CBR method in software fault prediction. We have assessed similarity between values of attributes. We employ the indigenous tool to facilitate analogy, using product and process metrics as independent variable. We have also shown the programmer's ability or efficiency of the programmers on the basis of experience (see Table 1). In this paper, we have used five similarity functions, namely Manhattan, Canberra, Clark, exponential, and Euclidean distance. Error level calculations with respect to LOC and development time have been calculated, as well as exact/near matching case is retrieved from the knowledgebase. As per the result, we choose to reuse the solution that has been retrieved by Euclidean and exponential method, because similarities between values of attributes are more similar to the problem. It can be seen as a snapshot (see snapshot 1 through 5) as well as Table 2. Therefore, we can conclude that CBR is used as a method for solving the new problem with the help of past solutions. Thus, we can say that CBR is a methodology for solving problems.

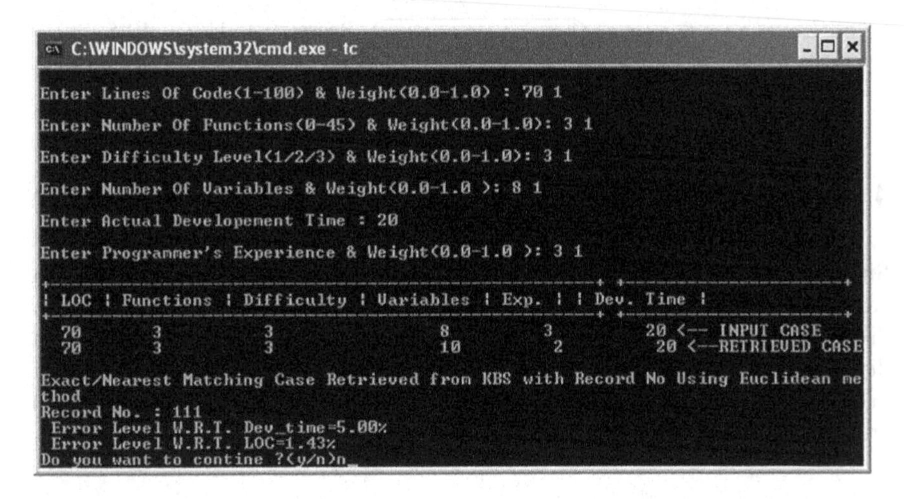

Snapshot 1

```
C:\WINDOWS\system32\cmd.exe - tc

Enter Lines Of Code(1-100) & Weight(0.0-1.0) : 70 1
Enter Number Of Functions(0-45) & Weight(0.0-1.0): 3 1
Enter Difficulty Level(1/2/3) & Weight(0.0-1.0): 3 1
Enter Number of Variables & Weight(0.0-1.0) : 8 1
Enter Actual Developement Time : 20
Enter Programmer's Experience & Weight(0.0-1.0) : 3 1
+------------------------------------------------------------+ +----------------+
! LOC ! Functions ! Difficulty ! Variables ! Exp. ! ! Dev. Time !
+------------------------------------------------------------+ +----------------+
  70        3            3               8          3           20    <-- INPUT
  68        3            2               8          2           20    <-- RETRIEVED
Exact/Nearest Matching Case Retrieved from KBS with Record No Using Manhattan me
thod
Record No. : 108
 Error Level W.R.T. Dev_time=20.00%
 Error Level W.R.T. LOC=5.71%
Do you want to contine ?(y/n)n
```

Snapshot 2

Table 2 Similarity functions used for overall error level calculation with respect to LOC and Dev_Time

Euclidean		Manhattan		Canberra		Clark		Exponential	
Error w.r.t. LOC (1.43 %)	Error w.r.t. Dev_Time (5 %)	Error w.r.t. LOC (5.71 %)	Error w.r.t. Dev_Time (20 %)	Error w.r.t. LOC (5.71 %)	Error w.r.t. Dev_Time (20 %)	Error w.r.t. LOC (11.43 %)	Error w.r.t. Dev_Time (40 %)	Error w.r.t. LOC (1.43 %)	Error w.r.t. Dev_Time (5 %)

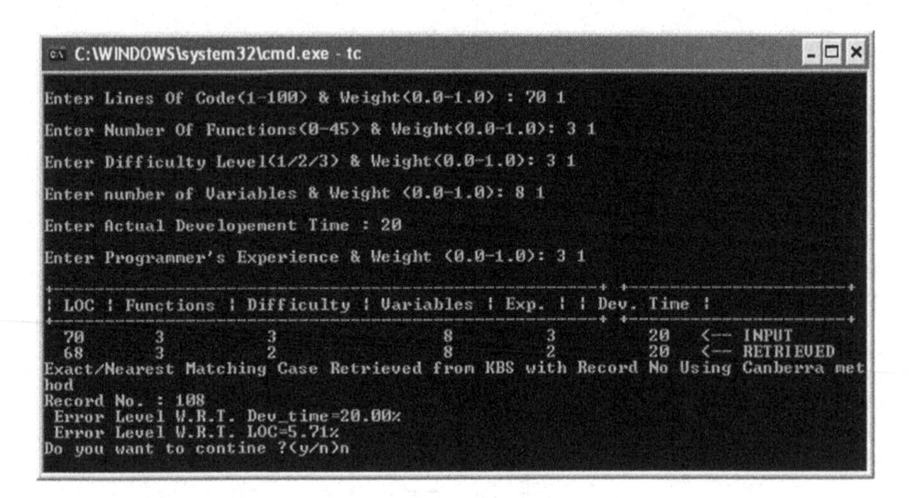

Snapshot 3

```
C:\WINDOWS\system32\cmd.exe - tc

Enter Lines Of Code(1-100) & Weight(0.0-1.0) : 70 1
Enter Number Of Functions(0-45) & Weight(0.0-1.0): 3 1
Enter Difficulty Level(1/2/3) & Weight(0.0-1.0): 3 1
Enter Number of Variables & Weight (0.0-1.0): 8 1
Enter Actual Developement Time : 20
Enter Programmer's Experience & Weight (0.0-1.0): 3 1
+-----------------------------------------------------------+ +---------------+
| LOC | Functions | Difficulty | Variables | Exp. | | Dev. Time |
+-----------------------------------------------------------+ +---------------+
  70       3           3              8          3         20    <-- INPUT
  65       2           1              8          3         20    <-- RETRIEVED
Exact/Nearest Matching Case Retrieved from KBS with Record No using Clark method

Record No. : 15
 Error Level W.R.T. Dev_time=40.00%
 Error Level W.R.T. LOC=11.43%
Do you want to contine ?(y/n)n_
```

Snapshot 4

```
C:\WINDOWS\system32\cmd.exe - tc
Enter Lines Of Code(1-100) & Weight(0.0-1.0) : 70 1
Enter Number Of Functions(0-45) & Weight(0.0-1.0): 3 1
Enter Difficulty Level(1/2/3) & Weight(0.0-1.0): 3 1
Enter Number of Variables & Weight (0.0-1.0): 8 1
Enter Actual Developement Time : 20
Enter Programmer's Experience & Weight (0.0-1.0): 3 1
+-------------------------------------------------------------+ +----------------+
| LOC | Functions | Difficulty | Variables | Exp. | | Dev. Time |
+-------------------------------------------------------------+ +----------------+
  70        3           3             8          3          20   <-- INPUT
  70        3           3             10         2          20   <-- RETRIEVED
Exact/Nearest Matching Case Retrieved from KBS with Record No using Exponential
method
Record No. : 111
 Error Level W.R.T. Dev_time=5.00%
 Error Level W.R.T. LOC=1.43%
Do you want to contine ?(y/n)n
```

Snapshot 5

References

1. Rashid, E., Patnaik, S., Bhattacherjee, V.: A survey in the area of machine learning and its application for software quality prediction. ACM SigSoft Softw. Eng. Notes **37**(5), (2012). doi:http://doi.acm.org/10.1145/2347696.2347709. ISSN 0163-5948 (New York)
2. Begum, S., Ahmed, M.U., Funk, P., Xiong, N., Folke, M.: Sch. of Innovation, Design and Engineering, Malardalen University, Vasteras, Sweden. IEEE Trans. Syst. Man, Cybern. C: Appl. Rev. **41**(4), (2011). doi:10.1109/TSMCC.2010.2071862. ISSN:1094-6977
3. Finnie, G.R., Sun, Z.: R_5 model for case-based reasoning. Knowl.-Based Syst. **16**(1), 59–65 (2003)
4. Watson, I.: Case-based reasoning is a methodology not a technology. Knowl.-Based Syst. **12,** 303–308 (1999)
5. Michael, M.M., Weber, R.O.: Case-based reasoning. A textbook, p. 546. Springer, Berlin (2013)
6. Rashid, E., Patnaik, S., Bhattacherjee, V.: Machine learning and software quality prediction: as an expert system. Int. J. Inf. Eng. Electron. Bus. **6**(2), 9–27 (2014). doi:10.5815/ijieeb.2014.02.02. ISSN:2074-9023 (Print), ISSN:2074-9031 (Online) [indexed in ISI, IET NSPEC. (Impact Factor 0.12)]
7. Rashid, E., Patnaik, S., Bhattacherjee, V.: Estimation and evaluation of change in software quality at a particular stage of software development. Indian J. Sci. Technol. **6**(10), (2013). ISSN:0974-5645 (Indexed in SCOPUS, ISI)
8. Challagulla, V.U.B. et al.: A unified framework for defect data analysis using the MBR technique. In: Proceeding of the 18th IEEE International Conference on Tools with Artificial Intelligence (ICTAI'06)

Various Strategies and Technical Aspects of Data Mining: A Theoretical Approach

Ekbal Rashid, Srikanta Patnaik and Arshad Usmani

Abstract In this paper, we are going to look at a very interesting aspect of database management, namely data mining and knowledge discovery. This area is attracting interest from not only researchers but also from the commercial world. The utility of data mining in commerce is more interesting than perhaps research areas. This has also raised many debates such as rights of privacy, legality and ethics, and rights to non-disclosure of information. It has someway opened a Pandora's box. Only, time will tell whether it is on the whole destructive or constructive; nonetheless, technology is not as such absolutely constructive or destructive; it only depends on how it is brought into use. In this paper, we have discussed about the technical aspects of data mining and what are the different strategies of data mining. Its sections give many technical aspects for various data mining and knowledge discovery methods, and we have given a rich array of examples and some are drawn from real-life applications.

Keywords Data mining · Classification · Algorithm · Clustering · Association

1 Introduction

Let us brief out first as to why is data mining necessary.

If one needs to talk to a manager or some Internet search on why data mining is necessary, there would be a variety of answers. Some would say it is very good for wealth generation, some would say it helps me to understand how market is

E. Rashid (✉) · A. Usmani
Cambridge Institute of Technology, Ranchi, India
e-mail: ekbalrashid2004@yahoo.com

A. Usmani
e-mail: ausmani@yahoo.co.in

S. Patnaik
SOA University, Bhubaneswar, Orissa, India
e-mail: patnaik_srikanta@yahoo.co.in

I.K. Sethi (ed.), *Computational Vision and Robotics*, Advances in Intelligent Systems and Computing 332, DOI 10.1007/978-81-322-2196-8_6

behaving, and some would say it would be great for security purposes. That is to identify abnormal activities in the network. This is an important area of data mining in these days. In short, there is no doubt about the growing influence of this aspect of data management.

The rest of the paper is structured as follows: Sect. 1 gives introduction, Sect. 2 describes the type of data. In Sect. 3 we present data mining and statistical inferences. In Sect. 4, association and item sets discussed. Section 5 presents classification and clustering with conclusion.

1.1 What Is Data Mining then?

It is the generic term used to look for hidden patterns and trends in data that is not apparent in just summarizing the data [1]. We can look at patterns of students' performance in certain subjects. Whether students performing well in a particular subject will do good in some other one. Putting in a very broad sense, data mining is controlled by something called 'interestingness criteria.' Finding something or finding everything according which one does not know about according to some criteria. We start with a database. We give some interestingness criteria and then discover trends. The output should be one or more patterns that we did not know to exist. In data mining, we do not talk about any data mining query. Rather, it is the data mining algorithm that should give us something that we do not know. When we talk about patterns, we say something is a pattern and something is not a pattern. For this, we have to see what kind of data we are looking at. We also have to keep in mind what is the type of interestingness criteria that we are looking at.

2 Types of Data

The different types of data that we encounter are [2]:

Tabular or relational database in the form of tables. This is the most common form of data. Then, there are special data, having coordinates and attributes. There are temporal data which are data with a time tag associated with it, like the traffic in a network or database activity logs. Then, tree data like xml databases, sequence data such as information about genes. There are text and multimedia data.

When we talk about interestingness criteria, we can talk about frequency, that is how often something happens, or rarity which may point toward abnormal behavior. Co-relations between data, consistency, can also be considered to be interestingness criteria. A customer comes once every month is more consistent than a customer who comes ten times first month and not even a single time the second month and so on. Periodicity may be another interestingness criterion.

3 Data Mining and Statistical Inferences

When we talk about data mining, there is sometimes a confusion that data mining is the same as statistical inferences. There is fundamental difference between the two. In statistical inference, we start with a null hypothesis; that is, we make a hypothesis about the system, like if exams are in March, then the turnout will be higher. Based on this, we do sampling and this is very important in statistical inference. Based on this sampling, we verify or refute the hypothesis. There is some questionnaire that knows what to look for.

In data mining, we just have a huge data set. We do not have any null hypothesis. We use some interestingness criteria to mine this data set. Usually, there is no sampling done. The data set is mined at least once to look for patterns. We present a data mining algorithm for the data set. Sometimes it is also called 'hypothesis discovery.'

4 Association and Item Sets

Two fundamental concepts which are of interest in data mining are of association and item sets [2] .We also have the concept of support and confidence. The support for a given rule R is the ratio of the number of occurrences of R given all occurrences of all rules. The confidence of a rule $x \rightarrow y$ is the ratio of the number of occurrences of y given x among all other occurrences given x. The apriori algorithm is used in data mining for frequent item sets. The apriori algorithm goes as such:

Given is a minimum support s as interestingness criteria.

Step 1. Search for all individual elements (1 element item set) that have a minimum support of s.

Step 2. REPEAT

a) From the results of i element item sets, search for the results of i + 1 element item sets that have a support of s.

b) This becomes the set of all frequent (i + 1) item sets that are interesting.

Step 3. UNTIL item set size reaches maximum.

The property of item sets is that you consider item sets as atomic; that is, there is no point of ordering in the item sets. As for example, it does not matter whether a customer buys item 1 first or item 2 first, as long as they are bought together. From this, we can draw the inference that the two items are quite likely to be brought together in one piece. How can this information be useful? Say if one has a supermarket and data mining has resulted in the understanding that two particular items are likely to be bought together in a single packet, such as bags and books are likely to be bought together, then from this, it would be wise for the supermarket owner to place bags and books side by side in the supermarket so that it would be easier for people to select both of them.

However, if we are looking toward association rules then there is a sense of direction in it. If we say if X then Y, it is different from saying if Y then X. Hence, association mining requires two different thresholds, the minimum support as in the item set and the minimum confidence with which we can determine that a given association rule is interesting. The following is the method to mine association rules using apriori.

1. First use apriori to generate different item sets of different sizes.
2. At each iteration, divide the item set into two parts the LHS and the RHS. This gives us a rule of the form LHS → RHS.
3. The confidence of this rule is that support of LHS → RHS divided by the support of LHS.
4. The confidence of all those rules are discarded which have a support of less than that of minimum confidence.

5 Classification and Clustering

Next, there are the tasks of discovering a classification or of discovering clusters within a data set. So come the concepts of classification and clustering. Intuitively, they seem to do the same thing. However, there is a marked difference between the two. Classification maps elements to one of the sets of predetermined classes based on the difference among data elements belonging to different classes [3].

Clustering groups data elements into different groups based on the similarity between elements within a single group [4]. In classification, we know the classes beforehand while mostly in clustering we do not know how many clusters we are going to get. In data mining, we are interested in discovering classification. For example, suppose we have data about cricket matches that have been played in a particular city. Now, this city is notorious for its frequent changes in weather. Suppose we have data such as if the day is sunny and the temperature is 30°, play is continued. If the day is overcast and temperature is 17°, then play is abandoned; if day is sunny and temperature is 20°, then still the play is continued, so on and so forth. Now, the classification problem is that whether one can classify the weather conditions, that is, how the day was and the temperature into one of the two classification criteria, whether play will be organized or it will be abandoned.

For classification problem, we use the Hunt's algorithm for decision tree identification. It goes as follows:

Given N different element types and m different decision classes,
Step 1. For i ← 1 to n do

a. **Add element i to the i-1 th element item sets from the previous iteration.**
b. **Identify the set of decision classes for each item set.**
c. **If an item set has only one decision class, then it is done, remove that item set from the subsequent iteration.**

Now, we look into the methods of clustering. Clustering is philosophically different from classification. Classification is the method of increasing the differences between the elements so as to make them belong to different classes, while clustering is the process of increasing the similarities between different elements so as to form them into different clusters. So clustering essentially partitions data sets into several clusters. What is the property of a particular cluster? The property is that the similarity of different elements in one cluster is much more than the similarity between different elements across clusters. So members belonging to the same cluster are much more similar to one another than they are to some members of some other cluster. And there are several measures of similarities and most of which are reduced to geometric similarities by projecting these data sets into hypercubes or n-dimensional spaces and then using some kind of distance measures such as Euclidean distance measures or Manhattan distance to compute the similarity. The first kind of clustering algorithm is called the nearest neighbor clustering algorithm. This clustering algorithm takes a parameter called threshold or the maximum distance t between the members of a given cluster [5].

Given n elements x1, x2,… xn and given a threshold t,

Step 1. j ← 1, k ← 1, cluster = []
Step 2. Repeat
find the nearest neighbor of xj
let the nearest neighbor be in some cluster m
if distance to nearest neighbor is greater than t, then create a new cluster and increment the number of clusters or assign it to the cluster m
j ← j-1
Step 3. until j > n

There is another kind of clustering technique which is also popular and it is called iterative conditional clustering. This differs from the nearest neighbor technique in the sense that here, the number of clusters are fixed beforehand. In the nearest neighbor technique, the numbers of clusters are not fixed beforehand. That means one does not know how many clusters one is going to get given a particular threshold and a data set.

Given n different elements and *k* different clusters and each with a center, this center means the centroid in the statistical sense.

Step 1. Assign each element to the closest cluster center
Step 2. After all assignments have been made, compute the cluster centoids for each of the cluster. This means one has to compute the average of all the points that have made up this cluster. Possibly, this will shift the centroid to a different location.
Step 3. Repeat the above two steps with the new centroids until the algorithm converges or stabilizes so that the centroids will stop shifting and then we know that we have found the best centroids for each of the clusters.

Iterative conditional clustering is essentially a technique where something like saying that suppose I have a data set, and suppose I want to create ten different clusters out of this data set, where would these clusters lie. On the other hand, nearest neighbor clustering technique would say suppose I have this data set and suppose I have some maximum distance, between elements that can lie between a data set, then how many clusters will I find. This concludes about technical aspects and various strategies of data mining and clustering.

References

1. Dunhum, M.H.,: Data Mining: Introductory and Advanced Topics, 3rd Impression. Pearson Education, India (2008)
2. Srinivasa, S.: Data Mining, Data Warehousing and Knowledge Discovery Basic Algorithms and Concepts. www.anilsoft.com
3. Han, J., Kamber, M.: Data Mining Concepts and Techniques, 2nd edn. Morgan Kaufmann Publishers, Massachusetts (2006)
4. Triantaphyllou, E.: Data Mining and Knowledge Discovery Approaches Based on Rule Induction Techniques. Springer, Berlin (2009). ISBN:1441941738 9781441941732
5. Catal, C., Sevim, U., Diri, B.: Clustering and metrics threshold based software fault prediction of Unlabeled Program Modules. In: 6th International Conference on Information Technology: New generations, pp. 199–204. IEEE (2009)

CBDF-Based Target Searching and Tracking Using Particle Swarm Optimization

Sanjeev Sharma, Chiranjib Sur, Anupam Shukla and Ritu Tiwari

Abstract Target tracking and searching are the very important problems in robotics. It can be used in many variations like path planning where the objective is to reach up to target without collide with obstacle, or it may be used in the places where some source is needed to find. Here nature inspired PSO algorithm is used to solve this problem with the help of multi robot system. Use of multi robots will find the target fast. Here new concept clustering based distributing factors (CBDF) is introduced to scatter the robots in environment to search and track the target. This CBDF method divides the area into regions. Different robots use coordination to reach up to these targets. For the movement of robots, PSO is used because it can be considered as minimization problem with respect to minimize the path length. Here, four parameters move, time, coverage, and energy are calculated to reach up to target. At last, we are showing the results for both known and unknown target problem.

Keywords Target tracking · Target searching · Particle swarm optimization · Clustering-based distribution factor

S. Sharma (✉) · C. Sur · A. Shukla · R. Tiwari
ABV-Indian Institute of Information Technology and Management, Gwalior, India
e-mail: sanjeev.sharma1868@gmail.com

C. Sur
e-mail: chiranjibsur@gmail.com

A. Shukla
e-mail: dranupamshukla@gmail.com

R. Tiwari
e-mail: tiwariritu2@gmail.com

I.K. Sethi (ed.), *Computational Vision and Robotics*, Advances in Intelligent Systems and Computing 332, DOI 10.1007/978-81-322-2196-8_7

1 Introduction

Target tracking and target searching are the fundamental problems in robotics, which have many applications for finding thesource where human is not able to work. Here, robots move in the environment and detect the target by sensors.

This problem can be solved by using single robot or multirobots. But obviously, work with the multirobot will be advantageous because of fast finding of target, and if some of the robots fail, then it is possible that target can be found by other, so it will be robust.

Swarm-based robotics [1–3] can work better in this situation because of cooperation and reaching the best which will be target. One of the most suitable algorithms is particle swarm optimization. Many literature are available that inspired for this work.

Sedigheh et al. [4] proposed a navigation method using PSO. This method worked for local navigation. First convert the navigation problem into optimization problem based on some target parameters. Then, PSO searches the optimal solution from the search space. An evaluation function is to be calculated for each particle. A global best is to be calculated in each iteration, and robot move to next point toward the goal. Here, robot has limited capability to avoid the obstacle based on the sensors. The results are tested for dynamically changing environment where the obstacle may be static.

Ellips et al. [5] presented a multiobjective PSO-based Algorithm for Robot path planning. Here, particle swarm optimization is used for global path planning while probabilistic road map is used for local that will avoid the obstacle. Two objective functions are used in PSO equation where smoothness is measured by difference of angles of hypothetical line that connect the robot with two successive positions to its goal. PRM generates the random node with the help of PSO; however, good PSO particle are added as a auxiliary node.

Dai el al. [6] described the robot path planning using rough set genetic algorithm. Author combined the genetic algorithm and rough set theory to optimize the path and to enhance the precision. Here, initial decision making table is to be obtained and then rough set theory is used to extract the rule. These rules will be used to train the initial population in genetics algorithm. The results show that using this optimizes path faster, as compared to simple GA.

Kala et al. [7] uses the hierarchical multineuron heuristic search and probability-based fitness for robot path planning. They used the A* star with heuristics to perform the better path planning. Here, each generation gives the better path that has higher probability of reaching. Probability approach is used for map representation. Here, each cell has probability 0 or 1, which defines the presence or absence of obstacle in cell. This algorithm gives the better result as compared to standard A*.

Ant colony optimization [8] is also used for robot path planning. Here, each ant drops a pheromone that marks the point that has been visited. Based on this pheromone value, the next ant heuristically determines its next movement based on its equation on probability of selecting a path. In this manner, all ants guided to global target.

Rashmi et al. [9] presented the multirobot path planning using honey bee mating optimization (HBMO) algorithm. Here, objective is to select the ultimate shortest path without collide with the obstacle. HBMO used here for local path for individual robots. Here, for finding the better position of robot, two objective functions are constructed, first one is to decide the next position of the robot, and second one to avoid the collision with obstacle and other robot. Here, two metrics are used to measure the performance of algorithm, one is average total path deviation and second one is average uncovered target distance.

The rest of the paper is organized as follows: Section 2 discusses the basic particle swarm optimization algorithm. Section 3 discusses the modified algorithm that will work for robot movement. Clustering-based distribution factors are in Sect. 4. Flow graph is discussed in Sect. 5. In Sect. 6, experiments and results are given. At last, in Sect. 7, conclusion is given.

2 Classical Particle Swarm Optimization

PSO can be applied in many optimization problems because of its simple concept, unique searching mechanism, and computational intelligence. Each particle represents a solution. In each step, particle gets closer to optimum solution by its own behaviors and social behaviors of other particles. Each solution is represented by four vectors, its current position, best position found so far, the best position found by its neighborhood so far, and its velocity, and adjusts its position in the search space based on the best position reached by itself and on the best position reached by its neighborhood during the search process [10–15].

The position and velocity update formula for each particle is given by these equations.

$$x_i(t+1) = x_i(t) + v_i(t+1) \tag{1}$$

$$v_i(t+1) = \omega v_i(t) + c_1\varphi 1(P_{\text{ibest}} - x_i) + c_2\varphi 2\left(P_{\text{gbest}} - x_i\right) \tag{2}$$

where $\varphi 1$ and $\varphi 2$ are random variables, that is, uniformly distributed within [0, 1]. c_1 and c_2 are the weight change factors. ω is the inertia weight. P_{ibest} represents the best position of the particle, and P_{gbest} represents the best position of swarm.

These are the steps in PSO algorithm:

1. Initialize position and velocity of all particles randomly in N dimension space.
2. Evaluate the fitness of each particles
3. Compare the each individual particle fitness value with its pbest value. If the current value is better than its previous best, make this current value as pbest.
4. Find gbest among all the particles. This gbest has the best fitness value.
5. Update the velocity and position according to Eqs. (1) and (2).
6. Repeat Steps 2–5 until a stop criterion is satisfied or a predefined number of iterations are completed.

3 Modified Particle Swarm Optimization

For target tracking and searching, some modifications are needed. That can be suggested or considered as described below.

A big concern is the global best and the inertia best, and more importantly, they do not exist as the solution is not complete and it is very difficult to determine any temporary fitness regarding it. But we can actually or up to a certain extent determine the fitness function approximately using the remaining distance for each agent left to reach the target when the target is known. So less the distance left better is the particle, but in reality, in constrained environment, it is half true and there are much more than it is anticipated.

However, in case of target searching, what we did is determination of the area coverage made by an agent and more the area covered more is its fitness. In this way, the positional vector is mapped with the fitness vector of the PSO particle. Rest of the algorithm goes as it is traditionally and as the value of c_1 and c_2 decides the significant amount to be taken from each term, its value needs to be determined efficiently.

4 Directional Movement

Clustering-based distribution factor is the new concept which guides the robot in all direction to search the target. In clustering-based distribution factor, the whole region is divided into a number of regions say x and from any position of the workspace, we can have x direction if we consider that the region head or cluster head is the final point and the position of the agent as the initial point. Using this phenomenon, the robots get the directionality factor and a way to get out of the local region or local optima. There are two kinds of clustering-based distribution factor.

4.1 Scheme 1—Directional Scattering Effect

Directional Scattering Effect (DSE) is provided to guide the robot to go pass a certain explored area and when there is requirement of changing a region, come out of a trap or an enclosure, etc. The direction factor is derived out of the cluster head and normally selected randomly to explore new direction and outings.

4.2 Scheme 2—Zig-Zag Search Effect

Zig-Zag Search Effect (ZSE) is like the DSE in all respect but mostly used in local search. Here, the direction changes more frequently and thus helping more hovering over the workspace. Here, also the cluster head-based direction is chosen dynamically and randomly.

Figures 1 and 2 show the diagram for DSE and zig zag effect.

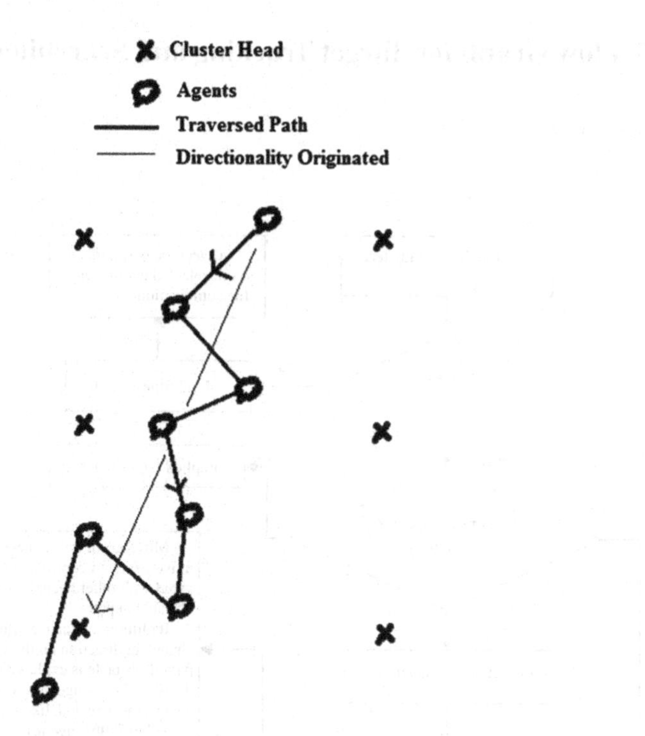

Fig. 1 Directional scattering effect

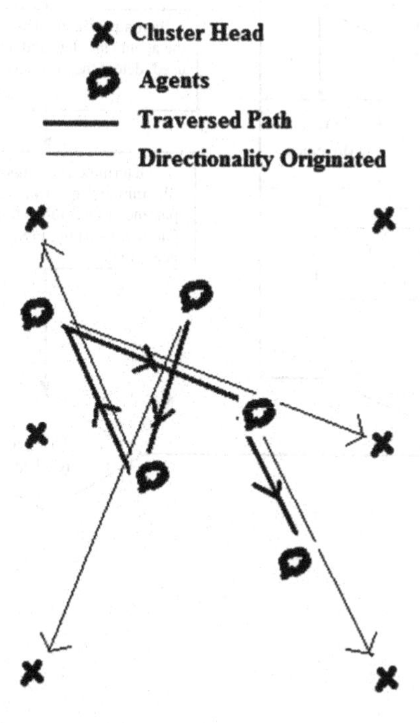

Fig. 2 Zig-zag search effect

5 Flow Graph for Target Tracking and Searching

See Fig. 3.

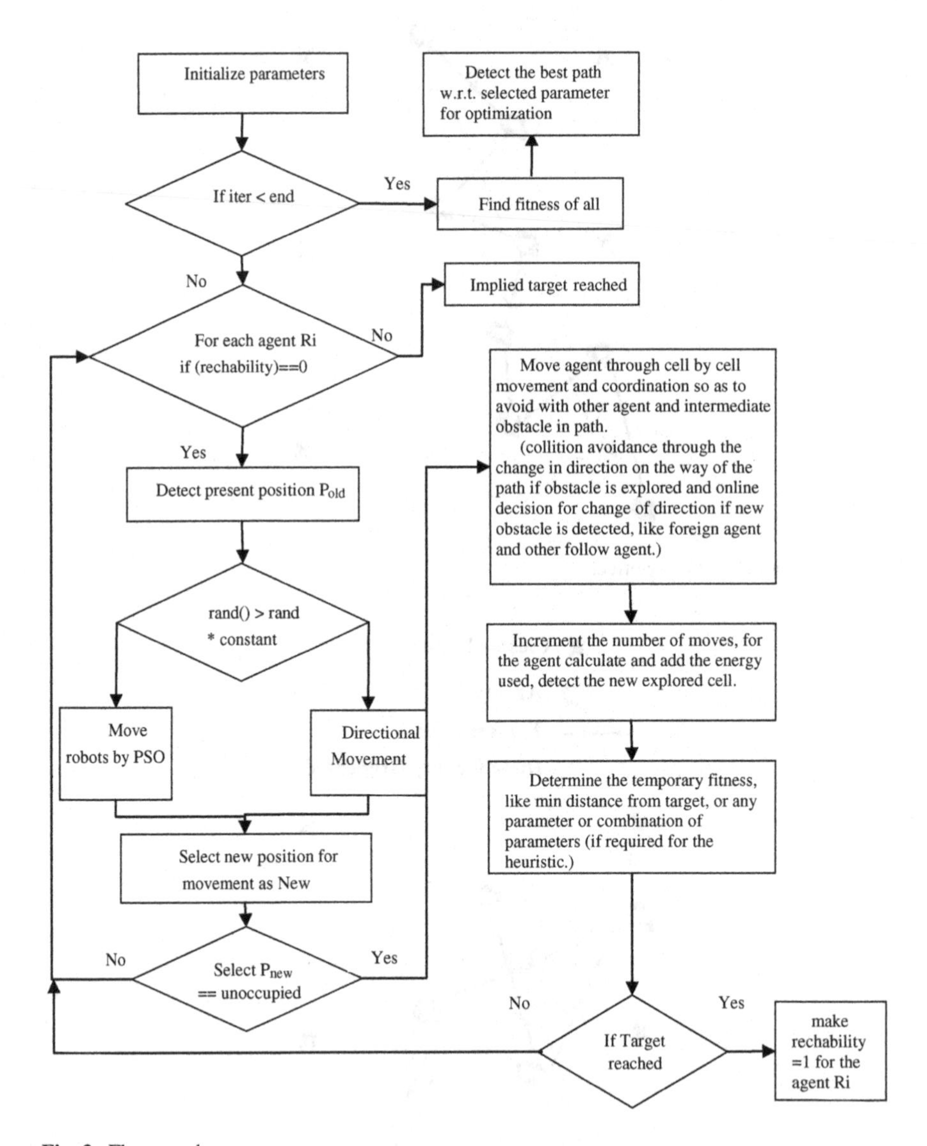

Fig. 3 Flow graph

6 Experiments and Results

Here, the problem is classified into two cases. One is where the target is known or we can say that robot has direction to reach up to target. Another is target is unknown where the robot does not have any idea so move into any direction to reach up to target.

6.1 Environment Descriptions

Here, two environments are considered, one is multiobstacle and another is chess-like that are given in Fig. 4.

6.2 Parameters Setup

See Table 1.

6.3 Path—Unknown Target

Here, we are showing the traverse path to reach up to target from source. For movement from one point to another, we are using specified PSO algorithm.

Here, robots find the paths that are guided by PSO algorithm. Figure 5a shows the result for multiobstacle environment. Figure 5b shows the result for chess environment.

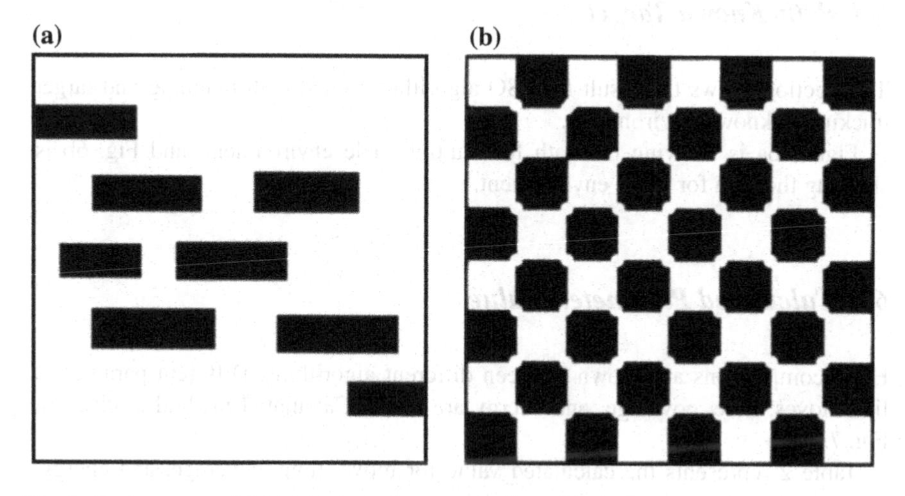

Fig. 4 Environment. **a** Multiobstacle environment. **b** Chess-like environment

Table 1 Experimental Parameters

Parameters	Value
Map size	600 × 600
Sensor range	8
Target	Known/Unknown
Number of robots	5
Start point	Any point set as start

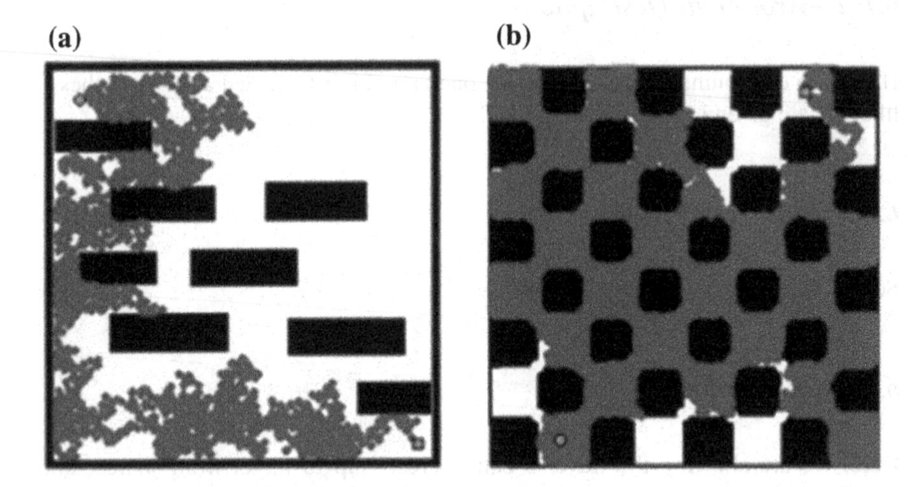

Fig. 5 Explored area to reach up to target for unknown target. **a** Multiobstacle environment. **b** Chess-like environment

6.4 Path-Known Target

This section shows the result for PSO algorithm guided path planning and target tracking in known environment.

Figure 6a is showing the path for multiobstacle environment, and Fig. 6b is showing the path for chess environment.

6.5 Calculated Parameters Value

Here, comparisons are shown between different algorithms. Different parameters like moves, time, coverage, and energy are taken. Calculated method is given in Fig. 7.

Table 2 represents the calculated value for move, time, coverage, and energy, respectively.

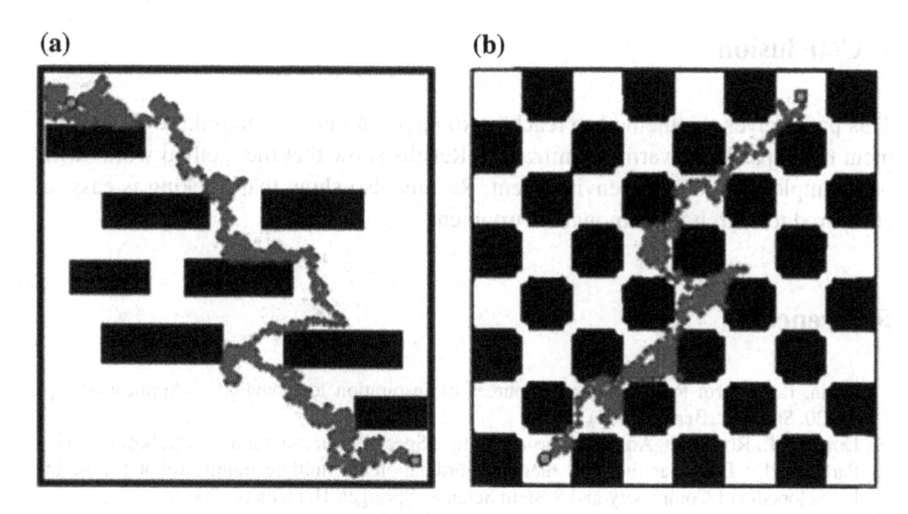

Fig. 6 Explored area to reach up to target for known target. **a** Multiobstacle. **b** Chess-like environment

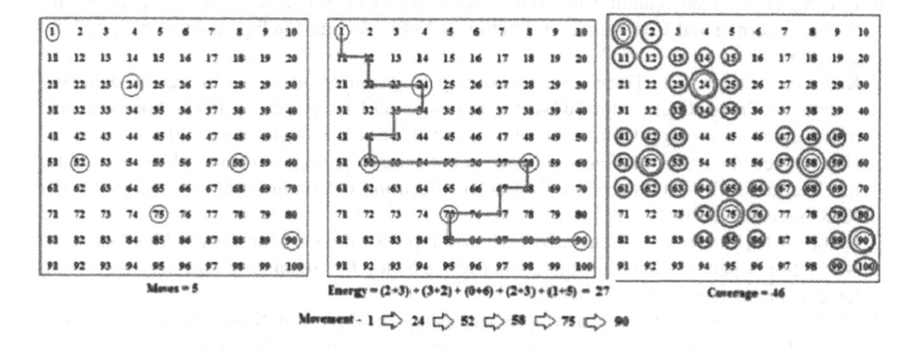

Fig. 7 Calculated parameters

Table 2 Calculated value

	Map1		Map2	
Parameters	Known target	Unknown target	Known target	Unknown target
Move	8,849	10,138	5,090	5,246
Time	1.83879	11.6628	1.075406	5.3072
Coverage	38,428	32,294	19,508	21,369
Energy	50,911	4,635	28,144	16,795

7 Conclusion

This paper gives the method to reach up to target in known and unknown environment using particle swarm optimization. Results show that the method works with both simple and complex environment. Results also show that tracking is easy as compared to search in unknown environment.

References

1. Şahin, E.: Swarm Robotics: From Sources of Inspiration to Domains of Application, pp. 10–20. Springer, Berlin (2005)
2. Dorigo, M., Roosevelt, A.F.D.: Swarm Robotics. Special Issue, Autonomous Robots (2004)
3. Parker, L.E.: Path planning and motion coordination in multiple mobile robot teams. In: Encyclopedia of Complexity and System Science. Springer, Heidelberg (2009)
4. Ahmadzadeh, S., Ghanavati, M.: Navigation of mobile robot using the PSO particle swarm optimization. J. Acad. Appl. Stud. (JAAS) **2**(1), 32–38 (2012)
5. Masehian, E., Sedighizadeh, D.: A multi-objective PSO-based algorithm for robot path planning. IEEE International Conference on Industrial Technology (ICIT). IEEE (2010)
6. Dai, S., et al.: Path planning for mobile robot based on rough set genetic algorithm. In: Second International Conference on Intelligent Networks and Intelligent Systems, pp. 278–281 (2009)
7. Kala, R., Shukla, A., Tiwari, R.: Robotic path planning using multi neuron heuristic search. In: Proceedings of the ACM 2nd International Conference on Interaction Sciences: Information Technology, Culture and Human, pp. 1318–1323. Seoul, Korea(2009)
8. Han, Q., et al.: Path planning of mobile robot based on improved ant colony algorithm. In: International Conference on Consumer Electronics, Communications and Networks (CECNet). IEEE (2011)
9. Sahoo, R.R., Rakshit, P., Haidar, M.T., Swarnalipi, S., Balabantaray, B.K., Mohapatra, S.: Navigational path planning of multi robot using honey bee mating optimization algorithm. Int. J. Comput. Appl. **27**(p1), (2011)
10. Eberhart, R.C., Kennedy, J.: A new optimizer using particle swarm theory. In: Proceedings of the Sixth International Symposium on Micro Machine and Human Science, vol. 1, pp. 39–43 (1995)
11. Pugh, J., Martinoli, A.: Inspiring and modeling multi-robot search with particle swarm optimization. In: Proceeding in Swarm Intelligence IEEE Symposium, SIS, pp. 332–339 (2007)
12. Zhang, Y., Gong, D.W., Zhang, J.H.: Robot path planning in uncertain environment using multi-objective particle swarm optimization. Elesvier J. Neurocomput. **103**, 172–185 (2013)
13. Saska, M., Macas, M., Preucil, L., Lhotská, L.: Robot path planning using particle swarm optimization of Ferguson splines. In: IEEE Conference on Emerging Technologies and Factory Automation, ETFA, pp. 833–839 (2006)
14. Van den Bergh, F., Engelbrecht, A.P.: A study of particle swarm optimization particle trajectories. Inf. Sci. **176**(8), 937–971 (2006)
15. Akhmedova, S., Semenkin, E.: Co-operation of biology related algorithms. In: IEEE Congress on Evolutionary Computation (CEC), pp. 2207–2214 (2013)

Real-Time Noise Canceller Using Modified Sigmoid Function RLS Algorithm

V.K. Gupta, D.K. Gupta and Mahesh Chandra

Abstract In this paper, modified sigmoid function RLS (MSRLS) algorithm is proposed for online noise cancellation from audio signals. The experiments are performed using TMS320C6713 processor with code composer studio (CCS) v3.1. The performance of RLS and MSRLS algorithms is evaluated and compared for noisy signals with car noise, F16 noise, and babble noise at −5, 0, and 5 dB SNR levels. The proposed MSRLS algorithm has shown a maximum of 2.03 dB improvement in SNR over RLS algorithm at input signal of −5 dB SNR with F16 noise. The proposed MSRLS algorithm has also shown decrement in mean square error (MSE) at all SNR levels for all noises in comparison with RLS algorithm.

Keywords Adaptive filter · RLS · MSRLS · TMS320C6713 · SNR · MSE

1 Introduction

The problem of controlling the noise level in the environment has been the focus of a tremendous amount of research over the years. In the process of transmission of information with a microphone, through a channel, noise automatically gets embedded to the signal which affects the quality of the audio signal. The objective is noise minimization and quality improvement of the signal through hardware

V.K. Gupta (✉)
Department of ECE, Inderprastha Engineering College, Ghaziabad, India
e-mail: guptavk76@gmail.com

D.K. Gupta
Department of ECE, Krishna Engineering College, Ghaziabad, India
e-mail: deepak_gpt@rediffmail.com

M. Chandra
Department of ECE, Birla Institute of Technology, Mesra, Ranchi 835215, India
e-mail: shrotriya69@rediffmail.com

I.K. Sethi (ed.), *Computational Vision and Robotics*, Advances in Intelligent Systems and Computing 332, DOI 10.1007/978-81-322-2196-8_8

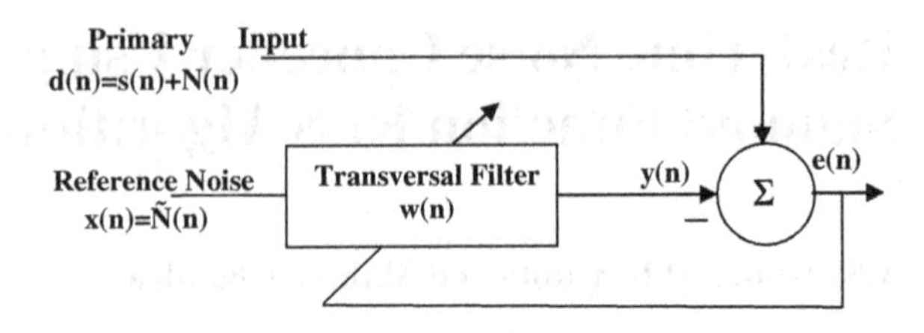

Fig. 1 Block diagram of an adaptive noise canceller

implementation of adaptive noise canceller (ANC) [1, 2]. The input to ANC is audio signal that is corrupted with self-generated background noise, and the designed digital filter reduces the noise level from the corrupted signal. Noise minimization is essential for speech signal transmission, audio signal processing, and reception due to ever-growing applications of telephone and cellular communication.

Adaptive algorithms play an important role in many diverse applications such as communications, acoustics, speech, radar, sonar, seismology, and biomedical engineering [3, 4]. There are basically two types of adaptive filter algorithms: least mean square (LMS) and recursive least squares (RLS) algorithms. Most of the other algorithms are evolved over the years from these two basic algorithms. Low computational complexity and simple structure are main advantages of LMS algorithm, but its rate of convergence is slow. RLS adaptive algorithm [5] with fast convergence is unlike the LMS algorithm convergence which is sensitive to array of parameters related to the input signal. In this paper, RLS adaptive filter algorithm and its variants are implemented on DSP processor TMS320C6713 [6] for noise minimization from audio signals. DSP techniques have been successful because of the development of low-cost software and hardware support. DSP processors are concerned primarily with real-time signal processing. Real-time processing requires the processing to keep pace with some external event, whereas non-real-time processing has no such timing constraint. DSP-based systems are less affected by environmental conditions. Code composer studio (CCS) provides an integrated development environment (IDE) to incorporate the software tools. CCS includes tools for code generation, such as a C compiler, an assembler, and a linker. It has graphical capabilities and supports real-time debugging. It provides an easy-to-use software tool to build and debug programs. After building and debugging the program, it is loaded to the DSP processors and executed.

The general setup of an ANC [7] is shown in the Fig. 1. Section 2 provides introduction of adaptive filters and details of RLS and MSRLS algorithms. Experimental setup is given in Sect. 3, and results are discussed in Sect. 4. Finally, Conclusions are drawn in Sect. 5.

2 Adaptive Filters

Adaptive filters continuously change their impulse response in order to satisfy the given conditions, and by doing so, change the very characteristic of their response. The aim of an adaptive filter is to calculate the difference between the desired signal and the adaptive filter output. This error signal is fed back into the adaptive filter, and its coefficients are changed algorithmically in order to minimize the cost function. When the adaptive filter output approaches to the desired signal, the error signal goes to zero. In this situation, the noise from the noisy signal will be minimized and the far user would not feel any disturbance.

2.1 RLS Algorithm

RLS algorithms are known for excellent performance when working in time-varying environments [5, 8]. This algorithm attempts to minimize the cost function shown in Eq. (1). The parameter λ is an exponential weighting factor that should be chosen in the range $0 < \lambda < 1$. This parameter is also called forgetting factor since the information of the distant past has an increasingly negligible effect on the coefficient updating. Hence, this algorithm gives more emphasis on recent input samples and tends to forget the past input samples. The cost function is expressed as follows:

$$\xi(n) = \sum_{k=1}^{n} \lambda^{n-k} e_n^2(k) \tag{1}$$

where $k = 1$ is the time at which the RLS algorithm starts. Unlike the LMS algorithm and its derivatives, the RLS algorithm directly considers the values of previous error estimations.

2.2 Proposed Algorithm (MSRLS)

In RLS algorithm, the forgetting factor remains constant. If the forgetting factor is made to vary as per the variability of noise in the signal, then RLS algorithm may perform better for noise minimization from the noisy signal. The variable forgetting factor was considered to improve the performance of the RLS algorithm [9–12]. The forgetting factor enhances the tracking capability and decreases the estimated error. The expression of the function [9] for variable forgetting factor is given in Eq. (2).

$$\lambda(t) = \lambda_{\min} + (1 - \lambda_{\min})\text{VFF} \tag{2}$$

where VFF is variable forgetting factor function used to convert forgetting factor λ from constant to variable value. The VFF used in this paper is modified sigmoid function as given in Eq. (3)

$$p(t) = \frac{1}{1 + a * \exp(-b * t)} \tag{3}$$

In modified sigmoid function, the sigmoid function is modified by using weight coefficients a and b which controls the variation of $p(t)$ which in turn controls the variations of λ .The values of a and b are taken in such a way that the variation of λ is always less than 1. In modified sigmoid RLS algorithm (MSRLS), λ varies as per Eq. (2) and the VFF function is taken from Eq. (3).

3 Experimental Setup

One DSK (DSP Starter Kit) board, one HCL laptop with CCS v3.1, one DELL PC with Goldwave software, and one Headphone (make Intex) were used in the experimental setup as shown in Fig. 3. The hardware implementation of ANC requires two inputs—the noisy signal and the noise. The TMS320C6713 DSK has two 32-bit input ports—MIC IN and LINE IN—and two output ports—LINE OUT and HEADPHONE OUT. Both input ports are multiplexed; that is, only one can work at a time. So for giving two inputs, stereo capability of the onboard AIC23 codec is used. One adapter with two inputs and one output is used for feeding noisy signal and noise to the LINE IN port of DSK as shown in Fig. 2. Noisy signal is given to left 16-bit channel, and noise is given to the right 16-bit channel of the adapter. Goldwave software is used for recording the output from LINE OUT port of DSK.

One clean sentence "YAHA SAI LAGHBAG PANCH MEAL DAKSHIN PASCHIM MAI KATGHAR GAON HAI" from Hindi speech database [13] has been taken as test sample. The noisy version of this sentence was prepared by adding car noise, F-16 noise, and babble noise from NOISEX-92 database [14] to this clean sentence at 0, −5, and 5 dB SNR levels using MATLAB 7.0. Both noisy sentence and noise are played repeatedly through Windows Media Player in PC and laptop. The noisy signal from PC and noise from laptop are given to LINE IN port of DSK through 2:1 adapter. The output of the ANC is played online through headphone connected to HEADPHONE OUT port of DSK. The output of the

Fig. 2 Experimental setup

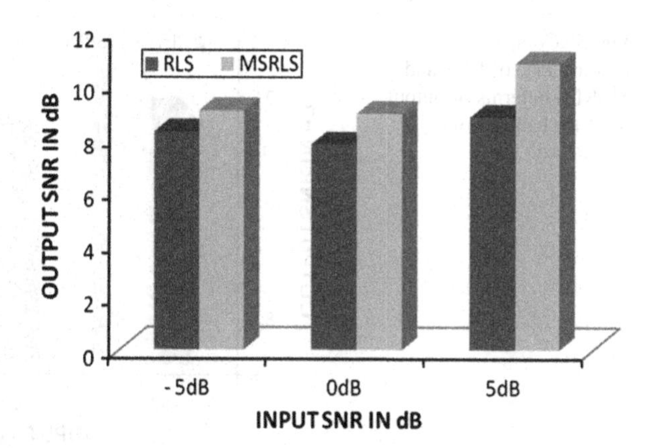

Fig. 3 Comparative performance of RLS and MSRLS in terms of output SNR for car noise

ANC is also given from LINE OUT port of DSK to audio input port of PC for recording through Goldwave software. The recorded output is saved in .wav file for calculation of output SNR and MSE.

4 Results and Discussion

Noisy data at −5 dB input SNR level corrupted with car noise and car noise were given to ANC through adapter. The value of forgetting factor λ for RLS was taken 0.99. The value of forgetting factor λ for MSRLS was calculated by using Eq. (2) by taking 0.95 value of λ_{min}. The programs of RLS and MSRLS were implemented on DSK using code composer studio. The output from HEADPHONE OUT port was played for online observations for listeners. The output from LINE OUT port was recorded for calculation of output SNR and MSE. The values of SNR and MSE were calculated in MATLAB 7.0 as explained in Sect. 3. Similarly, the results for both RLS and MSRLS algorithms were calculated for all three noises at all SNR levels. Figures 3, 4, and 5 show the improvement in SNR of

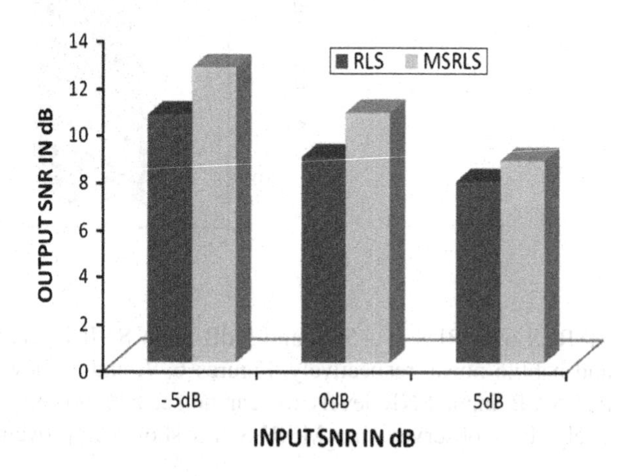

Fig. 4 Comparative performance of RLS and MSRLS in terms of output SNR for F16 noise

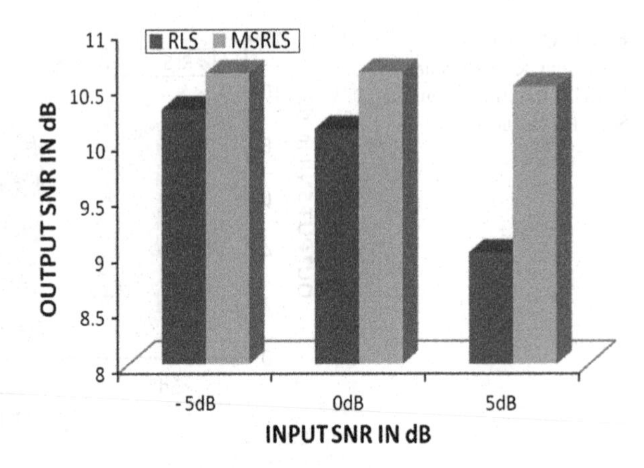

Fig. 5 Comparative performance of RLS and MSRLS in terms of output SNR for babble noise

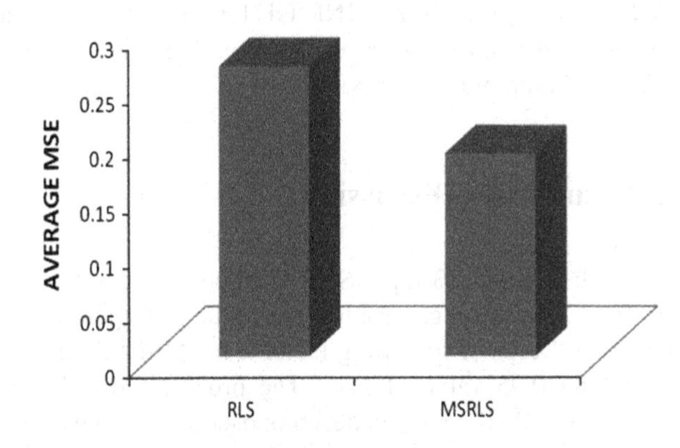

Fig. 6 Comparative performance of RLS and MSRLS in terms of average MSE of all three orders for car noise

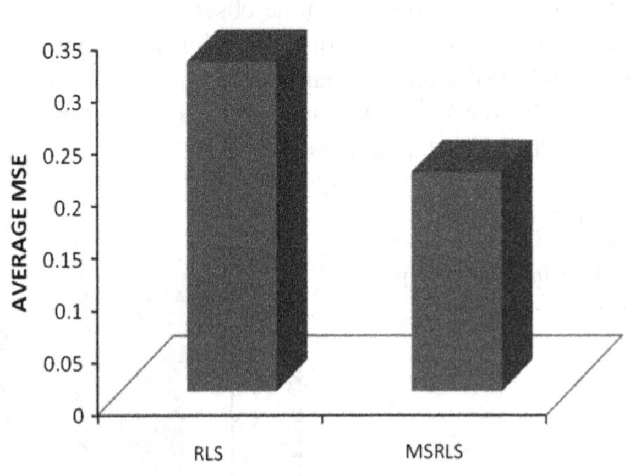

Fig. 7 Comparative performance of RLS and MSRLS in terms of average MSE of all three orders for F16 noise

MSRLS over RLS at −5, 0, and 5 dB input SNR levels for car noise, F16 noise, and babble noise, respectively. Figures 6, 7, and 8 show average MSE of −5, 0, and 5 dB input SNR levels for car noise, F16 noise, and babble noise, respectively. It is observed that MSRLS has shown improvements in SNR over RLS

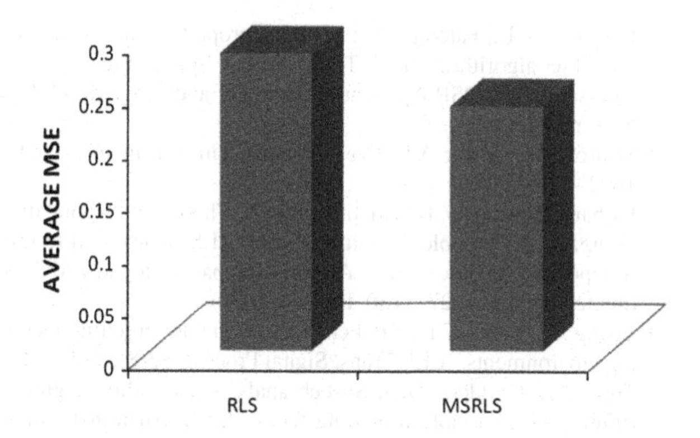

Fig. 8 Comparative performance of RLS and MSRLS in terms of average MSE of all three orders for babble noise

for all noises at all SNR levels. It is also observed that MSRLS has shown decrement in MSE over RLS for all noises at all SNR levels. This is due to the fact that SNR increment means increase in signal power which in turn decreases the MSE between filtered signal and clean signal. It is observed that MSRLS has shown a maximum of 2.03 dB improvement in SNR over RLS at −5 dB input SNR level for F16 noise. It is also observed that the improvement in SNR decreases when we move from lower-SNR level signal to higher-SNR level signals for all noises. As observed, RLS and MSRLS algorithms have shown a maximum of 15.47 dB and 17.5 dB improvements, respectively, in SNR for F16 noise at −5 dB SNR level.

5 Conclusion

The proposed MSRLS algorithm has shown its superiority over RLS algorithm for noise minimization for all noises at all SNR levels. It is due to the fact that the use of variable forgetting factor is able to track the variations in noisy signal more closely as compared to RLS algorithm. The factors a and b used in proposed algorithm are able to control the variations in forgetting factor more effectively for noise minimization.

References

1. Ryu, B.-S., et al.: Hardware implementation of an adaptive noise canceller in an automobile environment. In: Advanced Packaging and Systems Symposium, pp. 45–48. Seoul (2008)
2. Ryu, B.-S., et al.:The performance of an adaptive noise canceller with DSP processor. In: 40th Southeastern Symposium on System Theory, pp. 42–45. University of New Orleans, LA (2008)
3. Kuo, S.M., Morgan, D.R.: Active noise control: a tutorial review. Proc. IEEE **87**(6), 943–973 (1999)
4. Kuo, S.M., Kong, X., Gan, W.S.: Applications of adaptive feedback active noise control system. IEEE Trans. Contr. Syst. Technol. **11**(2), 216–220 (2000)

5. Eleftheriou, E., Falconer, D.: Tracking properties and steady-state performance of RLS adaptive fillter algorithms. IEEE Trans. Acoust. Speech Signal Process. **34**(5), 1097–1109 (1986)
6. Chassaing, R.: DSP Application Using C and the TMS320C6x DSK. Wiley Inter Science, New York (2002)
7. Widrow, B., et al.: Adaptive noise canceling: principles and applications. Proc. IEEE **63**, 1692–1716 (1975)
8. Farhang-Boroujeny, B.: Adaptive Filters, Theory and Applications. Wiley, New York (1999)
9. Wang, J.: A variable forgetting factor RLS adaptive filtering algorithm. In: International Symposium on Microwave, Antenna, propagation and EMC technologies for wireless communication, pp. 1127–1130. Beijing (2009)
10. Leung, S.-H., So, C.F.: Gradient-based variable forgetting factor RLS algorithm in time-varying environments. IEEE Trans. Signal Process. **53**(8), 3141–3150 (2005)
11. Ting, Y.T., Childers, D.G: Speech analysis using the weighted recursive least squares algorithm with a variable forgetting factor. In: International Conference on Acoustics, Speech, and Signal Processing. vol. 1, pp. 389–392. Albuquerque (1990)
12. Leung, S.H., So, C.F.: Nonlinear RLS algorithm using variable forgetting factor in mixture noise. In: International Conference on Acoustics, Speech, and Signal Processing, vol. 6, pp. 3777–3780. Salt Lake City (2001)
13. Samudravijaya, K., et al.: Hindi speech database. In: Proceedings of International Conference on Spoken Language processing (ICSLP00), Beijing (2000) (CDROM 00192.pdf)
14. Varga, A., Steeneken, H.J.M., Jones, D.: The NOISEX-92 study on the effect of additive noise on automatic speech recognition system. Reports of NATO Research Study Group (RSG.10) (1992)

Multiscale Local Binary Patterns for Facial Expression-Based Human Emotion Recognition

Swati Nigam and Ashish Khare

Abstract Facial expression is an important cue for emotion recognition in human behavior analysis. In this work, we have improved the recognition accuracy of facial expression recognition and presented a system framework. The framework consists of three modules: image processing, facial features extraction, and facial expression recognition. The face preprocessing component is implemented by cropping the facial area from images. The detected face is downsampled by bilinear interpolation to reduce the feature extraction area and to enhance execution time. For extraction of local motion-based facial features, we have used rotation-invariant uniform local binary patterns (LBP). A hierarchical multiscale approach has been adopted for computation of LBP. The selected features were fed into a well-designed tree-based multiclass SVM classifier with one-versus-all architecture. The system is trained and tested with benchmark dataset from JAFFE facial expression database. The experimental results of the proposed techniques are effective toward facial expression recognition and outperform other methods.

Keywords Human emotion recognition · Facial expression · Multiscale LBP · Multiclass SVM classifier

1 Introduction

Facial expressions provide important cues for human emotions and behavior analysis. Automatic recognition of facial expressions has been an interesting topic of research that has attracted much attention in the past few years. It has potential

S. Nigam (✉) · A. Khare
Department of Electronics and Communication, University of Allahabad, Allahabad, India
e-mail: swatinigam.au@gmail.com

A. Khare
e-mail: ashishkhare@hotmail.com

I.K. Sethi (ed.), *Computational Vision and Robotics*, Advances in Intelligent Systems and Computing 332, DOI 10.1007/978-81-322-2196-8_9

applications in several areas such as human–computer interaction, biometrics, data-driven animation, and customized applications for consumer products [1, 2]. However, different illumination conditions, pose, alignment, and occlusion problems make facial expression recognition problem much more difficult [3, 4]. Therefore, an automatic facial expression recognition system must be capable of handling these problems.

Several techniques have been proposed for facial expression recognition in the last decade. Recently, local binary patterns (LBP) and their variant-based facial expression recognition systems have gained much attention due to their better performance [5]. The LBP-based approaches are theoretically simple yet efficient. The LBP was originally a texture operator that has been used for face and facial expression recognition later [6]. The LBP retrieves local information by thresholding neighborhood values of the center pixel. It provides a method which is computationally efficient and robust toward monotonic changes in illumination. However, simple LBP has some limitations. It is not well capable of handling random noise and only considers the sign of the difference between two gray values and neglects the actual magnitude of the difference which is very important information [7, 8]. In order to handle these issues, local ternary patterns (LTP) [9] and local directional patterns (LDP) [10] have been proposed. In addition to these techniques, some more techniques have also been explored in the literature such as Gabor wavelets [11, 12] and local feature analysis [13]. A hybrid approach of local feature analysis-based method and image vector analysis-based method was proposed in [14].

However, the above mentioned approaches are not useful in real-time conditions. Therefore, in this work, we have presented a framework that consists of three modules: face preprocessing, facial features extraction, and facial expression recognition. Here, we have presented a hierarchical multiscale LBP-based technique that takes advantage of extracting information from different scales. This hierarchical multiscale LBP-based technique is efficient because it does not only consider uniform patterns at a time but also consider non-uniform patterns of different scales. The results have been tested over benchmark dataset, the JAFFE [15] facial expression database. Comparison with other methods in terms of recognition rate has shown better performance of the proposed method.

The rest of the paper is organized as follows: Sect. 2 describes image processing step of the proposed method, Sect. 3 briefly explains the proposed hierarchical multiscale LBP-based feature extraction technique, Sect. 4 deals with the classification strategy and experimental results, and discussions are given in Sect. 5, and Sect. 6 finally concludes this study.

2 Image Processing

All facial expression database images have been gone through a two-phase preprocessing scheme. The face detection part has been implemented by cropping the facial area of the images. Figure 1 shows sample face detection by the cropping of images.

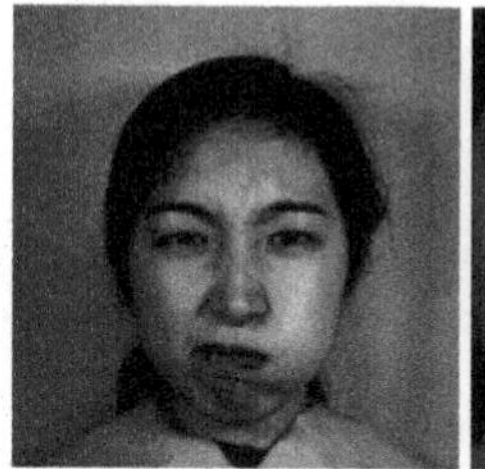 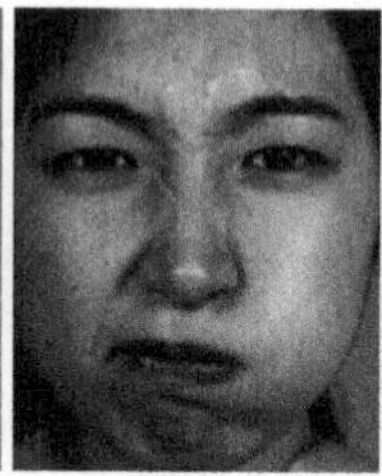 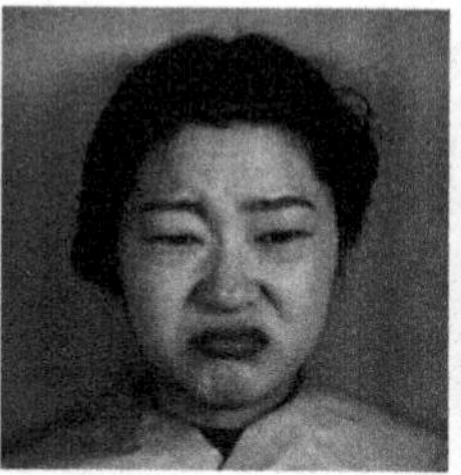 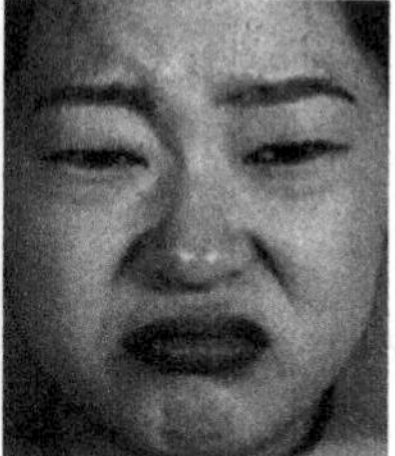

Fig. 1 Sample and cropped images from JAFFE database

The size of the images is scale normalized to a window of size 50 × 50 pixels. Most of the approaches have used bilinear interpolation for this purpose. We have also used the same technique. In this way, downsampling of facial images has been performed to reduce computation time.

3 Multiscale LBP

A LBP feature $\mathrm{LBP}_{P,R}$, of radius R and the intensities of the P sample pixel points, is given as below

$$\mathrm{LBP}_{P,R}(x,y) = \sum_{P=0}^{P-1} s(g_P - g_c)2^P, \tag{1}$$

where

$$s(u) = \begin{Bmatrix} 1, & u \geq 0 \\ 0, & u < 0 \end{Bmatrix}. \tag{2}$$

To remove the effect of rotation, the modified version with rotation invariance is defined as follows

$$\mathrm{LBP}_{P,R}^{\mathrm{ri}}(x,y) = \min\{\mathrm{ROR}(\mathrm{LBP}_{P,R}, i) | i = 0, 1, \ldots, P-1\} \tag{3}$$

where $\mathrm{ROR}(\mathrm{LBP}_{P,R}, i)$ performs a circular bitwise right shift on the P-bit number $\mathrm{LBP}_{P,R}$ for i times.

The uniform LBP are those LBP which have very few spatial transitions. Formally, uniform LBP have maximum two circular transitions between 0 and 1. For example, patterns 00000001 and 11111011 have only one and two transitions between 0 and 1, respectively; therefore, they are uniform patterns. Similarly, patterns 11110100 and 00001010 have 3 and 4 transitions between 0 and 1, respectively; therefore, they are not uniform LBP. The grayscale and rotation-invariant uniform LBP is defined as

$$\mathrm{LBP}_{P,R}^{\mathrm{riu2}} = \begin{cases} \sum_{P=0}^{P-1} s(g_P - g_C), & \text{if } U(\mathrm{LBP}_{P,R}) \leq 2 \\ P+1, & \text{otherwise} \end{cases} \tag{4}$$

where

$$U(\mathrm{LBP}_{P,R}) = |s(g_{P-1} - g_C) - s(g_0 - g_C)| + \sum_{P=1}^{P-1} |s(g_P - g_C) - s(g_{P-1} - g_C)| \tag{5}$$

In the presented method, we have proposed to build a mutliscale LBP histogram from bigger radius to smaller radius. First, LBP map of the biggest radius for each pixel is built. The pixels are divided into two groups of uniform and non-uniform patterns. A subhistogram is built for those uniform patterns. Those pixels, whose pattern is non-uniform, are further processed to extract their LBP patterns by smaller radius. The process stops for pixels whose new patterns are uniform, and the remaining pixels are continued to extract LBP patterns by smaller radius until the smallest radius.

In the proposed hierarchical multiscale LBP scheme, the LBP histogram for $R = 4$ is first built. For those non-uniform patterns by $R = 4$ operator, a new histogram is built by $R = 3$ operator. Then, the non-uniform patterns of $R = 3$ are further proceeded to build a histogram by $R = 2$ operator. Finally, three histograms are concatenated into one multiscale histogram.

4 SVM Classification

The theory of the SVM classifier was first proposed by Cortes and Vapnik [16] in 1995. Recently, it has demonstrated good performance in two class classification problems [17, 18]. Therefore, it has become a topic of interest in machine learning for activity recognition too.

The OVA architecture is an initial implementation of the SVM classification for multiple classes. It requires same number of classifiers as number of classes. All samples are considered simultaneously when training each classifier. In this classification technique, one class is kept at positive label and all other classes are kept at negative label at a time. This process goes on for each class. For the N-class problems ($N > 2$), N 2-class SVM classifiers are constructed. The ith SVM is trained while labeling all the samples in the ith class as positive examples and the rest as negative examples. In the recognition phase, a test example is presented to all N SVMs and is labeled according to the maximum output among the N classifiers. In several examples, it is also proved that OVA architecture shows least error rate as compared to other architectures.

5 Experiments and Results

In this experiment, we demonstrate the robustness of the proposed method for different facial expressions. For this purpose, we have selected the Japanese Female Facial Expression (JAFFE) dataset [15]. The database contains 213 images of 7 facial expressions (6 basic facial expressions and 1 neutral) posed by 10 Japanese female models. Each image has been rated on 6 emotion adjectives by 60 Japanese subjects. These results have been shown in Table 1.

In Table 1, we can observe that the proposed method shows 100 % recognition rate for the expressions fear, happy, neutral, sad, and surprised. For the expression angry, we get a high rate of 96.67 %. Only in case of expression disgust, we get a no recognition rate. The average recognition rate is 85.2 %. Therefore, we can say that the proposed method provides a high recognition rate for facial expression recognition in JAFFE database. Processing time of whole database is 54.4 s. Comparison of the proposed method with other existing methods in terms of percentage recognition rate is shown in Table 2.

It is clear from Table 2 that the proposed method performs better in comparison with other existing methods in terms of percentage recognition rate.

Table 1 Confusion matrix of 7-class facial expression recognition over JAFFE

	Angry (%)	Disgust (%)	Fear (%)	Happy (%)	Neutral (%)	Sad (%)	Surprised (%)
Angry	96.67	3.33	0	0	0	0	0
Disgust	0	0	0	0	0	0	100.00
Fear	0	0	100.00	0	0	0	0
Happy	0	0	0	100.00	0	0	0
Neutral	0	0	0	0	100.00	0	0
Sad	0	0	0	0	0	100.00	0
Surprised	0	0	0	0	0	0	100.00

Table 2 Comparison of recognition rates of different methods over JAFFE

Authors	Approach used	Database	Recognition rate (%)
Kumbhar et al. [13]	Image features	JAFFE	60–70
Shinohara and Otsu [14]	Local features	JAFFE	69.4
Proposed method	Local features	JAFFE	85.2

6 Conclusions

This study employs an advanced technique to improve the recognition rate as well as computation time of facial expression recognition methods. Face preprocessing has been carried out by cropping of facial expressions area. These faces were downsampled by the interpolation-based scale resizing. A hierarchical multiscale approach has been adopted to extract uniform rotation-invariant uniform LBP of facial images. The approach reduced the image dimensions and preserved the perceptual quality of the original images. The extracted features were fed into a multiclass support vector machine classifier with one-versus-all architecture. The classifier was trained with sample dataset from the JAFFE facial expression recognition database. The proposed method was compared with several methods, and the performance was found to be better than those methods.

Acknowledgments This work was supported by the Council of Scientific and Industrial Research, Human Resource Development Group, India, via Grant No. 09/001/(0362)/2012/EMR-I.

References

1. Tian, Y.L., Kanade, T., Cohn, J.F.: Facial expression analysis. In: Handbook of Face Recognition, pp. 247–275. Springer, New York (2005)
2. Pantic, M., Rothkrantz, L.J.M.: Automatic analysis of facial expressions: the state of the art. IEEE Trans. Pattern Anal. Mach. Intell. **22**(12), 1424–1445 (2000)
3. Bettadapura, V.: Face Expression Recognition and Analysis: The State of the Art. Tech report (2012). http://arXiv:1203.6722
4. Fasel, B., Luettin, J.: Automatic facial expression analysis: a survey. Pattern Recogn. **36**(1), 259–275 (2003)
5. Shan, C., Gong, S., McOwan, P.: Facial expression recognition based on local binary patterns: a comprehensive study. Image Vis. Comput. **27**(6), 803–816 (2009)
6. Pietikäinen, M.: Computer Vision Using Local Binary Patterns, vol. 40. Springer, Berlin (2011)
7. Ahmed, F., Bari, H., Hossain, E.: Person-independent facial expression recognition based on compound local binary pattern (CLBP). Int. Arab J. Inf. Technol. **11**(2), 195–203 (2014)
8. Luo, Y., Wu, C.M., Zhang, Y.: Facial expression recognition based on fusion feature of PCA and LBP with SVM. Optik-Int. J. Light Electron. Optics **124**(17), 2767–2770 (2013)
9. Ahmed F., Kabir M.: Directional ternary pattern for facial expression recognition. In: Proceedings of the IEEE International Conference on Consumer Electronics, pp. 265–266. Las Vegas (2012)
10. Kabir, H., Jabid, T., Chae, O.: Local directional pattern variance: a robust feature descriptor for facial expression recognition. Int. Arab J. Inf. Technol. **9**(4), 382–391 (2012)
11. Lekshmi, V.P., Sasikumar, M.: Analysis of facial expression using Gabor and SVM. Int. J. Recent Trends Eng. **2** (2009)
12. Owusu, E., Zhan, Y., Mao, Q.R.: A neural-AdaBoost based facial expression recognition system. Expert Syst. Appl. **41**(7), 3383–3390 (2014)
13. Kumbhar, M., Jadhav, A., Patil, M.: Facial expression recognition based on image feature. Int. J. Comput. Commun. Eng. **1**(2), 117–119 (2012)

14. Shinohara, Y., Otsu, N.: Facial expression recognition using Fisher weight maps. In: Proceedings of IEEE International Conference Automatic Face and Gesture Recognition, pp. 499–504 (2004)
15. Dailey, M.N., Joyce, C., Lyons, M.J., Kamachi, M., Ishi, H., Gyoba, J., Cottrell, G.W.: Evidence and a computational explanation of cultural differences in facial expression recognition. Emotion **10**(6), 874 (2010)
16. Cortes, C., Vapnik, V.: Support vector networks. Mach. Learn. **20**(3), 273–297 (1995)
17. Nigam, S., Khare, A.: Multi-resolution approach for multiple human detection using moments and local binary patterns. Multimedia Tools Appl. (2014). doi:10.1007/s11042-014-1951-0 (Springer, published online on 3 April 2014)
18. Binh, N.T., Nigam, S., Khare, A.: Towards classification based human activity recognition in video sequences. In: Proceedings of 2nd International Conference on Context Aware Systems and Applications, pp. 209–218. Phu Quoc, Vietnam 25–26 Nov (2013)

Development of a New Algorithm Based on SVD for Image Watermarking

Arun Kumar Ray, Sabyasachi Padhihary, Prasanta Kumar Patra and Mihir Narayan Mohanty

Abstract The research on watermarking has been increasing day-by-day since past decade. It has been largely driven by its important applications in digital copyrights management and protection. To provide more watermarks and to minimize the distortion of the watermarked image, a novel technique is presented in this paper. In this paper, the singular value decomposition (SVD)-based image watermarking scheme is proposed. The output result of SVD is more secure and robust. SVD is often used to develop robust watermarking algorithms. However, the SVD-based algorithms exhibit false-positive problem and pose security concern. In this work, we try to overcome this problem. In the proposed schemes, the host image is first decomposed into sub-bands by applying discrete wavelet transform (DWT). The watermark image is embedded in all the sub-bands by modifying the singular values of each sub-band. Next to it, we propose to encrypt and embed the singular values of the watermark image instead of original singular values. RSA algorithm has been used for the encryption process. Peak signal-to-noise ratio (PSNR) is used to measure the imperceptibility of the proposed schemes. The simulation result shows its efficacy.

Keywords Watermarking · DWT · SVD · RSA algorithm · Encryption

A.K. Ray · S. Padhihary · P.K. Patra
School of Electronics Engineering, KIIT University,
Bhubaneswar, Odisha, India

M.N. Mohanty (✉)
School of Electronics Engineering, ITER, SOA University,
Bhubaneswar, Odisha, India
e-mail: mihirmohanty@soauniversity.ac.in

I.K. Sethi (ed.), *Computational Vision and Robotics*, Advances in Intelligent Systems and Computing 332, DOI 10.1007/978-81-322-2196-8_10

1 Introduction

Due to the advancement of digital multimedia tools, the storage and distribution of multimedia content become very easy. With the increase in the availability of digital data like multimedia services on the Internet, there is an impressed demand to manage and protect the illegal duplication of data. Digital watermarking is one of the solutions for such problem. It is to embed a secrete data (known as watermark) in the original data. Later, the watermark can be extracted and used for authentication and verification of ownership. The watermark may be a visual image or a code which is embedded in the host image as a digital signature of the owner so that it cannot be removed easily without affecting the host image. The watermarking schemes may be divided into two categories, such as transformed domain and spatial domain watermarking scheme. Many of the current techniques for embedding marks in digital images have been inspired by methods of image coding and compression. Digital watermarking methods have been recently proposed for various purposes and especially for copyright protection of multimedia data. With the development in multimedia technology and use of Internet, it has become a common practice to create copy, transmit, and distribute digital data. The digital watermarking system is an effective technique for protecting the copyright of the digital production and the data safety maintenance.

Transformed domain watermarking schemes produce better results than spatial domain watermarking scheme. Therefore, emphasis has been given to transformed domain watermarking schemes. The transformed domain watermarking schemes are based on the mathematical transformations such as discrete cosine transform (DCT) [1, 2], redundant discrete wavelet transform (RDWT) [3], discrete wavelet transform (DWT) [4–6], and fast Hadamard transform [4]. It also uses singular value decomposition (SVD) [3–13] and is widely used. Lots of researches have been done on the SVD-based watermarking schemes because of the stability of the singular values against different image processing attacks. Ganic and Eskicioglu [6] proposed a SVD-based watermarking scheme where the singular values of the watermark image were used to modify the singular values of the host image. Abdallah et al. [4] proposed a hybrid watermarking scheme comprising DWT, SVD, and FHM. These schemes are robust against image processing attacks but exhibit false-positive problem. Makbol et al. [3] proposed a robust watermarking scheme to solve this problem. According to the scheme, the watermark was directly embedded into the cover image. In this paper, we propose an improved SVD-based image watermarking scheme. Unlike other transforms which uses fixed orthogonal bases, SVD uses nonfixed orthogonal bases. The result of SVD gives good accuracy, good robustness, and good imperceptibility in resolving rightful ownership of watermarked image. The method is also suitable for noise and outliers removal, though other methods are there for outlier detection [14].

As robustness of watermarking scheme is of prime importance, we propose to use the features of human visual system (HVS) [11, 12]. Reddy and Varadarajan [12] proposed an efficient watermarking technique based on the entropy masking property of the HVS model. Additionally to improve the robustness, techniques of encryption along with the watermarking scheme were also used [13, 15]. In these schemes, chaotic maps were used to encrypt the watermark image. Then, the encrypted image was used in embedding process. Due to the high randomness of the encrypted sequence, the watermark image becomes more secure. This motivates us to use encryption technique in watermarking scheme. We use RSA algorithm for the encryption. To achieve high imperceptibility and robustness, we use the entropy masking property of HVS.

2 Proposed Watermarking Algorithm

In this work, we propose the watermarking scheme using SVD, but in a different approach. We embed the singular values of the watermark image in all the sub-bands of the host image. Instead of embedding the actual singular values, we embed the encrypted singular values so that the robustness increases. We implement RSA algorithm to encrypt the singular values.

DWT is used to select the areas of the host image in which a watermark can be embedded so that no significant change occurs to the visual quality of the host image. Applying 1-level DWT, the host image is decomposed into four sub-bands (such as LL, HL, LH, and HH) as shown in Fig. 1b. We use SVD because small change of the singular values does not affect the visibility of the host image. SVD decomposes an image into three matrices. Let's consider A be an image of size $M \times M$. The SVD of A given as $A = USV^T$, where U and V are orthogonal matrices, i.e., $U^T U = 1$ and $V^T V = 1$. S is a diagonal matrix and the diagonal values of S are called the singular values of A.

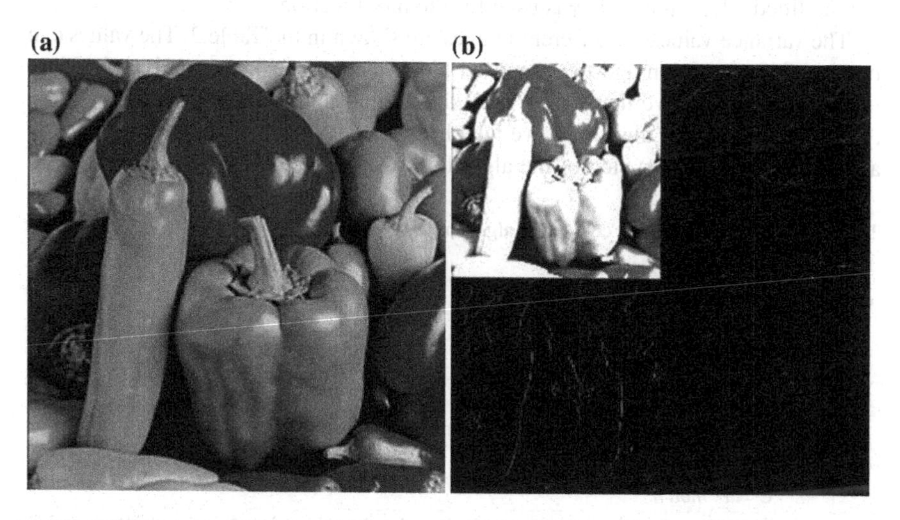

Fig. 1 DWT decomposition. **a** Pepper image of size 256 × 256. **b** 1-level Haar DWT image of size 128 × 128

2.1 Algorithm: Watermarking Scheme Based on RSA Algorithm

In this approach, the singular values of the watermark are encrypted using the concept of RSA algorithm. These encrypted singular values are used in embedding process. RSA algorithm is the most common public key algorithm which uses two keys for encryption and decryption, i.e., public key and private key. The data are encrypted with the help of public key, and at the destination, the encrypted data are decrypted with the help of private key. An attacker may know the public key, but without the help of private key, it becomes impossible to decrypt the original data. Generally, the keys used in the encryption and decryption vary from 64 to 1,024 bits long [1]. As a result, it becomes difficult for a hacker to hack the system. The algorithm for the encryption and decryption is shown in Table 1.

RSA algorithm has been used to modify the singular values of the watermark image. Because the singular values of different images do not vary much. From the result, it is clear that the variance of singular values of different images is significant after the encryption.

2.1.1 Watermark Embedding

In this section, the watermark embedding procedure is discussed. Before embedding the watermark image in the host image, we encrypt the singular values of the watermark image. To do so, we apply SVD to the watermark image. Usually, the singular values are real numbers. Therefore, we change the singular values to their integer form. Then using the RSA algorithm shown in Table 1, the singular values are modified. Here, public key is used for the modification.

The variance values for different images are shown in the Table 2. The values can be used for encryption and the proposed block diagram for watermark embedding is shown in Fig. 2. We develop an embedding algorithm which is shown in Table 3. For the embedding algorithm, we take host image 'A' and watermark image 'W' as our input. In the first step of the algorithm, we apply DWT to the host image to

Table 1 RSA encryption and decryption algorithm

Input	: Two prime numbers p, q
Output	: Public key e, n
	Private key d
1.	Find $n = P \times Q$.
2.	Find $\Phi\ (n) = (P - 1) \times (Q - 1)$.
3.	Select e such that $1 < e < \Phi\ (n)$ and e is co-prime to $\Phi\ (n)$.
4.	Find $d = e^{-1}\ mod\ \Phi\ (n)$.
5.	$C = P^e\ mod\ n$.
6.	$P = C^d\ mod\ n$, where e is the public key, d is the private key, P is the plaintext, and C is the ciphertext.

Table 2 Variance values

Images	Variance of singular values	
	Original image	Encrypted image
Peeper image	108.3121	35,980
Cameraman image	139.2399	34,454
Baboon image	140.0980	14,994

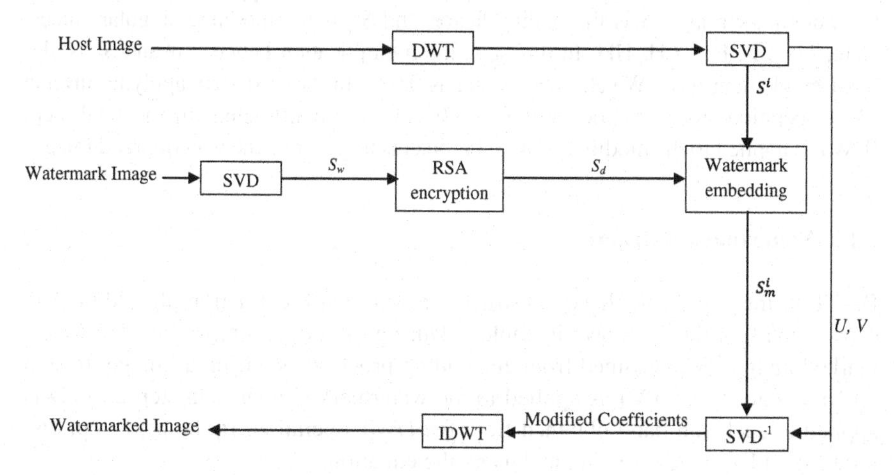

Fig. 2 Block diagram of watermark embedding for first algorithm

Table 3 Watermark embedding algorithm

Input	: Host image A,
	Watermark image W,
	Public key e
Output	: Watermarked image I
1.	Perform DWT to decompose the host image A into four sub-bands, A^i where $i = LL$, HL, LH, and HH.
2.	Apply SVD to each sub-band, $U^i S^i V^i = SVD(A^i)$.
3.	Apply SVD to the watermark image, $U_w S_w V_w = SVD(W)$.
4.	Encrypt the singular values of the watermark image W by applying RSA encryption, $S_d = (S_w)^e mod\, n$ (e = Public key).
5.	Modify the singular values of all the sub-bands with the encrypted singular values of the watermark image, $S^i_m = S^i + \alpha S_d$, where α = Scaling factor.
6.	Find the new modified DWT coefficient for each sub-band, i.e., $A^{i\prime} = U^i S^i_m (V^i)^T$.
7.	Perform IDWT using 4 sets of modified DWT coefficients to obtain the watermarked image, I.

decompose it into four sub-bands. In step 2, SVD is applied to all the sub-bands. In the next step, the singular values of the watermark image are obtained by applying SVD to it. In step 4, the singular values are encrypted using RSA algorithm as discussed above. In step 5, the encrypted singular values are used to modify the singular values of each sub-band. The modification is done by using the equation,

$$S_m^i = S^i + \alpha S_d \tag{1}$$

where S^i is the singular values of each sub-band, S_d is the encrypted singular values of the watermark image, α is the scaling factor, and S_m^i is the modified singular values. Here, i = LL, HL, LH, HH. In this step, the multiplication between α and S_d is element-by-element wise. We choose α value as 0.005. In the next step, applying inverse SVD operation, four modified DWT coefficients are constructed. In the final step, IDWT is applied to the modified DWT coefficients to generate the watermarked image.

2.1.2 Watermark Extraction

Based on the Fig. 3, we have constructed a watermark extraction algorithm. The extraction algorithm is shown in Table 4. During the extraction, we take the watermarked image 'W' (obtained from embedding process) as our input image. In step 1 of the algorithm, DWT is applied to the watermarked image. In step 2, SVD is applied to each sub-band obtained from the DWT operation. In the next step, distorted singular values are obtained using the equation,

$$S_n^i = \left(S^{i'} - S^i\right) / \alpha \tag{2}$$

where $S^{i'}$ is the singular values of each sub-band of the watermarked image, S^i is the singular values of each sub-band of the original image, and α is the scaling factor. We choose the same value of α as used in the embedding process. Here, the division operation occurs is element-by-element wise.

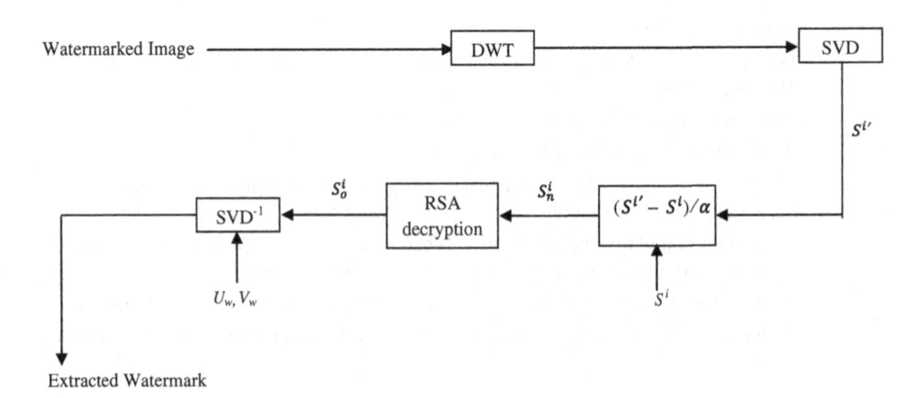

Fig. 3 Block diagram of watermark extracting for first algorithm

Table 4 : Watermark extraction algorithm

Input	: Watermarked image I
	Private key p
Output	: Extracted watermark image W
1.	Apply DWT to decompose the watermarked image I into 4 sub-bands, $I^{i'}$ where i = LL, HL, LH and HH.
2.	Apply SVD to each sub-band, i.e., $U^{i'}S^{i'}(V^{i'})^T = SVD\left(I^{i'}\right)$.
3.	Obtain the distorted singular values for each sub-band, $S_n^i = \left(S^{i'} - S^i\right)/\alpha$.
4.	Decrypt the distorted singular values, $S_o^i = (S_n^i)^p \, mod \, n$ (p = Private key).
5.	Extract the watermark image, $W^i = U_w S_o^i (V_w)^T$.

In step 4 of the extraction algorithm, the distorted singular values are decrypted using RSA algorithm as shown in Table 1. For the decryption process, we first rounded off the singular values. Using the private key generated from the RSA algorithm, the encrypted singular values are decrypted. Then, these decrypted singular values are changed to real numbers. Finally, using these singular values, required watermark image is extracted.

3 Simulation Results and Discussion

In this section, the experimental results are discussed. We carried out our experiment in MATLAB R2009b environment on a laptop with Intel i5 processor rated at 2.60 GHz and the operating system is Microsoft windows 7. For the experimental purpose, we use two grayscale images, i.e., one as the host image and another as the watermark image.

The quality of the watermarked image can be estimated with the help of a parameter known as peak signal-to-noise ratio (PSNR) and is expressed in decibel (dB).

The PSNR is given by,

$$\text{PSNR} = 10 \log \frac{255^2}{\text{MSE}} \text{ dB} \tag{3}$$

where A is the host image and A_w is the watermarked image of pixel value M. MSE is the mean square error between A and A_w.

3.1 Results

For this approach, we use pepper image as the host image of size 512 × 512 pixel (Fig. 4a) and cameraman image as the watermark image of size 256 × 256 pixel (Fig. 4b). The watermarked image is shown in Fig. 4c.

Fig. 4 First algorithm. **a** Host image, **b** Watermark image, and **c** Watermarked image

Fig. 5 Extracted watermark images from each sub-band for the first algorithm

From the experiment, we get a PSNR value of 27.3283 between the host image and the watermarked image. This implies that the proposed algorithm is sufficiently imperceptible and the embedded watermark image is not visible to human eye. We embed the watermark image in all sub-bands (i.e., LL, HL, LH, and HH). Therefore, after extraction, four watermark images are obtained. The extracted watermark images from each sub-band are shown in Fig. 5.

We also extract the watermark images from the attacked watermarked images. As the watermark image is embedded in all the sub-bands of the host image, four watermark images are extracted.

4 Conclusions

In this work, we proposed two image watermarking schemes. In both the schemes, we use RSA algorithm to encrypt the singular values of the watermark image. As a result, the variance of the encrypted singular values significantly changes. This increases the difficulty level for an attacker to extract the original singular values. Although we embed the watermark image in all the sub-bands, the watermark image is completely invisible in the watermarked image, which satisfies the imperceptibility requirement of the watermarking scheme. The PSNR value is obtained as 27.3823. The algorithm shows good robustness against the image processing attacks. Thus, we can say that the proposed algorithm is an efficient watermarking scheme.

References

1. Liu, F., Liu, Y.: A watermarking algorithm for digital image based on DCT and SVD. In: IEEE Congress on Image and Signal Processing, pp. 380–383 (2008)
2. Huang, F., Guan, Z.H.: A hybrid SVD–DCT watermarking method based on LPSNR. Pattern Recogn. Lett. **25**(15), 1769–1775 (2004)
3. Makbol, N.M., Khoo, B.E.: Robust blind image watermarking scheme based on redundant discrete wavelet transform and singular value decomposition. Int. J. Electron. Commun. (AEU) **67**(2), 102–112 (2013)
4. Abdallah, E.E., Hamza, A.B., Bhattacharya, P.: Improved image watermarking scheme using fast Hadamard and discrete wavelet transforms. J. Electron. Imaging **16**(3), 201–209 (2007)
5. Bhatnagar, G., Raman, B.: A new robust reference watermarking scheme based on DWT-SVD. Comput. Stand. Interfaces **31**(5), 1002–1013 (2009)
6. Ganic, E., Eskicioglu, A.M.: Robust DWT–SVD domain Image watermarking embedding data in all frequencies. In: Proceedings of the ACM Multimedia and Security workshop, pp. 166–174. Magdeburg, Germany (2004)
7. Liu, R., Tan, T.: An SVD-based watermarking scheme for protecting rightful ownership. IEEE Trans. Multimedia **4**(1), 121–128 (2002)
8. Ghazy, .R, El-Fishawy, N., Hadhoud, M., Dessouky, M., El-Samie, F.: An efficient block-by block SVD-based image watermarking scheme. In: Proceedings of the 24th National Radio Science Conference, pp. 1–9. Cairo, Egypt (2007)
9. Lai, C.C., Tsai, C.C.: Digital image watermarking using discrete wavelet transform and singular value decomposition. IEEE Trans. Instrum. Meas. **59**(11), 3060–3063 (2010)
10. Run, R.S., Horng, S.J., Lai, J.L., Kao, T.W., Chen, R.J.: An improved SVD-based watermarking technique for copyright protection. Expert Syst. Appl. **39**(1), 673–689 (2012)
11. Lai, C.C.: An improved SVD-based watermarking scheme using human visual characteristics. Opt. Commun. **284**(4), 938–944 (2011)
12. Reddy, V.P., Varadarajan, S.: An effective wavelet-based watermarking scheme using human visual system for scheme using human visual system for protecting copyrights of digital images. Int. J. Comput. Electr. Eng. **2**(1), 32–40 (2010)
13. Li, Y.L., Li, J.P., Ren, Q.B.: Based on chaotic encryption and SVD digital image watermarking. In: International Conference on Apperceiving Computing and Intelligence Analysis, pp. 285–289 (2010)
14. Dash, S., Mohanty, M.N.: Analysis of outliers in system identification using WLMS algorithm. In: IEEE Conference on Computing, Electronics and Electrical Technologies. ICCEET, Kanyakumari, 21–22 Mar 2012
15. Borie, J.C., Puech, W., Dumas, M.: Encrypted medical images for secure transfer. In: International Conference on Diagnostic Imaging and Analysis, pp. 250–255. ICDIA (2002)

Nonlinear System Identification Using Clonal Particle Swarm Optimization-based Functional Link Artificial Neural Network

Kumar Gaurav and Sudhansu Kumar Mishra

Abstract In this paper, the clonal particle swarm optimization (C-PSO)-based functional link artificial neural network model (FLANN) has been applied for the identification of a nonlinear system. System identification in different challenging situations such as noisy and time varying environments has been a matter of great concern for researchers and scientists for the last few decades. Other variants of the FLANN network, trained with some of the optimization techniques, such as the genetic algorithm (GA), particle swarm optimization (PSO), and comprehensive learning particle swarm optimization (CLPSO) have also been applied in this interesting field of research. The proposed C-PSO algorithm is based on the clonal principle in a natural immune system. In the C-PSO, the essence of the clonal operator is to generate a set of clone particles near the expected candidate solution. Hence, the search spaces are enlarged, and the diversity of the cloned particles is increased to avoid trapping in local minima. The simulation study reveals the superiority of the proposed C-PSO-based FLANN model, in terms of the convergence rate, over other competitive networks. The performance comparison is also carried out based on the computational complexity.

Keywords Functional link artificial neural network · Chebyshev polynomials · Nonlinear system identification · Clonal particle swarm optimization

K. Gaurav (✉) · S.K. Mishra
Birla Institute of Technology, Mesra, Ranchi, India
e-mail: gaurav8131@gmail.com

S.K. Mishra
e-mail: sudhansu.nit@gmail.com

I.K. Sethi (ed.), *Computational Vision and Robotics*, Advances in Intelligent Systems and Computing 332, DOI 10.1007/978-81-322-2196-8_11

1 Introduction

The Nonlinear system identification of a complex dynamic plant has many applications in many areas of the control theory, such as the power system, communication, and instrumentation. But the major concern of a complex dynamic plant is its identification. To perform these complex tasks in highly nonlinear environments, the artificial neural network proves to be a powerful technique [1]. This is because of the ability of the artificial neural network to learn, which is based on the optimization of an appropriate error function. Also ANNs give excellent performances for the approximation of nonlinear functions. However, the major drawback of an artificial neural network is that it requires a large computational complexity. Pao [1] proposed FLANN which can be used for functional approximation and pattern classification, with a faster convergence rate and lesser computational complexity than an MLP network. Sadegh [2] reported a functional basis perceptron network for functional identification and control of nonlinear systems. Narendra and Parthasarathy [3] proposed effective identification and control of a dynamic system using MLP networks. Patra et al. [4] showed that with the proper choice of functional expansion, FLANN performs better than the MLP in the system identification problems.

However, all these models use gradient search algorithms such as the least mean square (LMS), recursive least square (RLS), and back propagation (BP). The main disadvantage of these algorithms is that they may be trapped in local optimal point. To avoid this shortcoming, researchers have applied heuristic-based approaches such as the GA, PSO, and bacterial forging optimization (BFO) for training the neural network. All these heuristic techniques have been applied to different fields of applications in the last few decades. Mishra et al. [5, 6] have applied some of these evolutionary techniques in portfolio optimization problems. Because of the property of inherent parallelism, the GA improves its range of operation and the optima can be located more precisely. But the deficiency from which the GA mostly suffers is its speed, or the convergence rate with which it reaches to the global optima. So, the technique PSO was developed. Kennedy and Eberhart realized that an optimization problem can be formulated by mimicking the social behavior of a flock of birds flying across an area looking for food [7]. This observation and inspiration by the social behavior exhibited by flocks of birds and schools of fish resulted in the invention of a novel optimization technique called PSO. One variants of the PSO, named as CLPSO, has been introduced by Jiang and Qin [8], particularly aimed at overcoming some of the drawbacks of the PSO. This strategy enables the diversity of the swarm, chosen according to a specific logic, and helps to preserve the diversity within the swarm that discourages premature convergence. Lee and Jeng [9] reported the Chebyshev polynomial-based model of an ANN for static function approximation. It further showed that this FLANN has universal approximation capability and a faster convergence rate than an MLP network. Mishra et al. [10] showed that the FLANN having the Chebyshev functional expansion is better for salt and pepper noise suppression from a corrupted image. Mishra et al. [11] showed that the performance of FLANN structures with exponential function expansion is suitable for Gaussian noise suppression. Nanda et al. [12] concluded that the AIS-based model gives better results in the identification of nonlinear dynamic system. Nanda et al. [13] showed in their paper that the FLANN-AIS model gives superior performance over standard GA- and PSO-based approaches to the MIMO identification problem.

The organization of the present work is as follows: Sect. 2 introduces the identification problem. Section 3 deals with the FLANN model. Section 4 deals with the basic principle and the algorithm required for identification of the model, using CPSO is developed and represented. To validate the performance of the model, the simulation study of different nonlinear systems is carried out in Sect. 5. Finally, the conclusion of the proposed investigation is outlined in Sect. 6.

2 Identification of Nonlinear Systems

System identification is a fundamental problem in system theory, because it is the route to build a mathematical archetype for the anonymous system by monitoring its input output data. For a particular input, the prototype output should meet with its corresponding actual system output. The goal of the identification problem is to raise a relevant model that can generate an output $\hat{Y}$ which can approximate the plant output Y for the same input X, in such a way that the mean square error is the least.

3 Structure of an Artificial Neural Network

3.1 Multilayer perceptron

The MLP has a multilayer architecture with one or more hidden layers between its input and output layers. All the nodes of a lower layer are connected with all the nodes of the adjacent layer through a set of weights. All the nodes in all the layers (except the input layer) of the MLP contain a nonlinear tan*h*(.) function. A pattern is applied to the input layer without any computation. Thus, the output of the nodes of this layer is the input pattern itself. The weighted sum of the outputs of a lower layer is passed through the nonlinear function of a node in the upper layer to produce its output. Thus, the outputs of all the nodes of the network are computed. The outputs of the output layer are compared with a target pattern associated with the input pattern. The error between the target pattern and the output layer node is used to find least mean square (MSE), which is used as a cost function. The BP algorithm attempts to minimize the cost function by updating all the weights of the network.

3.2 Functional Link ANN

The FLANN, which was initially proposed by Pao, is a single layer artificial neural network structure, capable of performing complex decision regions by generating nonlinear decision boundaries. In a FLANN, the need for a hidden layer is removed. In contrast to the linear weighting of the input pattern produced by the linear links of a MLP, the functional link acts on the entire pattern by generating a set of linearly independent functions.

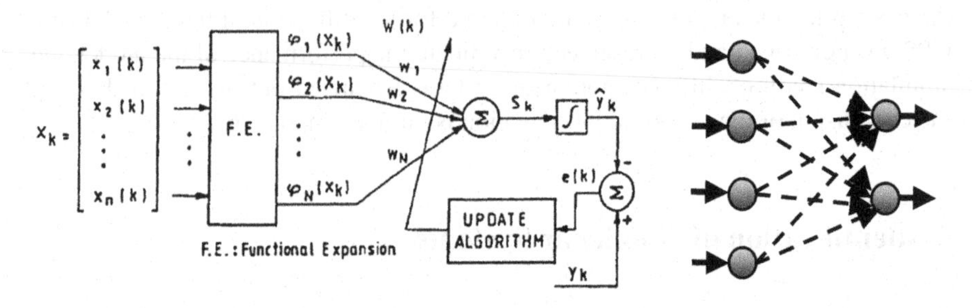

For the functional expansion of the input pattern, trigonometric, exponential, power series, the Chebyshev or any other functional expansion can be used. In this paper, the Chebyshev functional expansion is used and the structure of a Ch-FLANN is shown in figure below. Chebyshev polynomials are a set of orthogonal polynomials defined as the solution to the Chebyshev differential equation. These higher Chebyshev polynomials for $-1 < x < 1$ may be generated, using the mathematical series given by $T_{n+1} = 2xT_n(x) - T_{n-1}(x)$.

The first few Chebyshev polynomials are as follows:

$$\left.\begin{aligned} T_0(x) &= 1 \\ T_1(x) &= x \\ T_2(x) &= 2x^2 - 1 \\ T_3(x) &= 4x^3 - 3x \\ T_4(x) &= 8x^4 - 8x^2 + 1 \\ T_5(x) &= 16x^5 - 20x^3 + 5x \end{aligned}\right\}$$

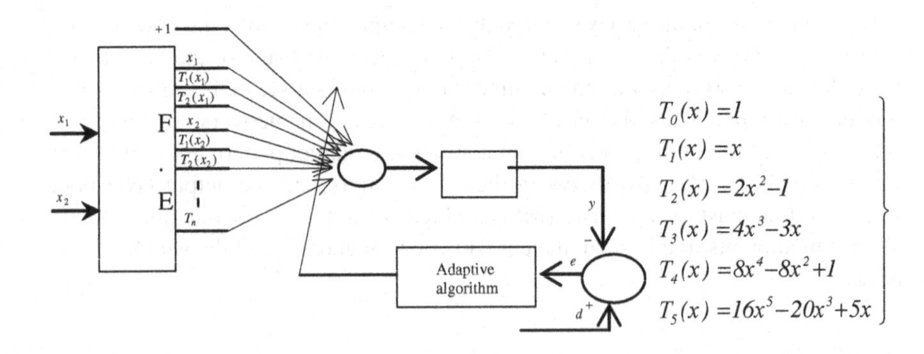

The architecture becomes simple, and training does not involve full BP. Thus, nonlinear modeling can be accomplished, by means of a linear learning rule, such as the delta rule. The computational complexity is also reduced and the neural net becomes suitable for online applications.

4 Basic Principle of the Clonal PSO

Tan et al. [14] introduced the clonal expansion process in a natural immune system (NIS) into the PSO, to strength the interaction between particles in a swarm and to improve the convergent performances of the standard PSO. According to the clonal expansion process in a NIS, a clonal operator is introduced by them. The clonal operator is used to copy one point as "*N*" same points according to its fitness function, and then generate "*N*" new particles by carrying out the mutation and selection processes. This is synonymous with the concentration mechanisms used for antigens and antibodies in NIS. In the CPSO also, definitions similar to those used in AIS have been implemented. Antigen, antibody, and the affinity between these two are correspond to the object optimization function, solution candidate, and the fitness value of the solution on the objective optimization function, respectively.

According to the PSO, the behavior of particle is being customized as follows:

$$v^{i,n+1} = w^n.v^{i,n} + \Phi_1 u_1 \times (p^i - x^{i,n}) + \Phi_2 u_2 \times (g - x^{i,n}) \quad (1)$$

$$x^{i,n+1} = x^{i,n} + v^{i,n+1} \quad (2)$$

4.1 Algorithm

Step 1. Initialization of parameter w^n, Φ_1, u_1, Φ_2 and u_2.
Step 2. The state evolution of particles is iteratively updated according to (1) and (2).
Step 3. Memory the global best fit particles of first Q generations as parent particle,

$$P^q_{gB}, \quad q = 1, 2, \ldots, Q \quad (3)$$

Step 4. Clone the memorized Q global best particles.
Step 5. We carry out the mutation operation for all the cloned particles. The Mutation process is implemented by using some random disturbances, i.e.,

$$P^{q^1}_{gB} = (1 - \mu)P^q_{gB} \quad (4)$$

Step 6. The current $p^{q^1}_{gB}$ is stored in memory. The other particles are selected according to a strategy of diversity by keeping the concentration mechanism. Hence, in the next generation of particles, a certain concentration of particles will be maintained for each fitness layer. The concentration of ith particles is defined as

$$D(x_i) = \left(\sum_{j=1}^{N+Q} \left| f(x_i) - f(x_j) \right| \right)^{-1}, \quad i = 1, 2, \ldots, N+Q \tag{5}$$

where x_i and $f(x_i)$ in Eq. (5) denote the ith particle and its fitness value, respectively.

The selection probability in terms of the concentration of particles can be derived as

$$p(x_i) = \frac{\frac{1}{D(x_i)}}{\sum_{j=1}^{N+Q} \frac{1}{D(x_j)}}, \quad i = 1, 2, \ldots, N+Q \tag{6}$$

It is clear from Eqs. (5) and (6) that the more the particles are similar to the antibody the less the probability of their selection and vice versa. Hence, the particle with a low fitness value also has an opportunity to evolve.

Step 7. Terminate the algorithm by considering some stop criteria. The stopping criteria may be a presetting accuracy of the solution or a given maximum number of generations. In the present study, the maximum number of generation has been taken as the stopping criteria.

5 Simulation Study

In this section, we carry out the simulation of all the variants of the FLANN model for the identification of the system. To validate our proposed algorithm, we have used an English nonlinear channel. While training, the amplitude of noise that affects the channel in different situations is taken as −30 to −10 dB. The configuration of the PC used for this identification is 2 GB RAM, 500 GB HDD, Intel core DUO processor, and the version of the Matlab used is MATLAB 7.10.0 (R2010a) (Table 1).

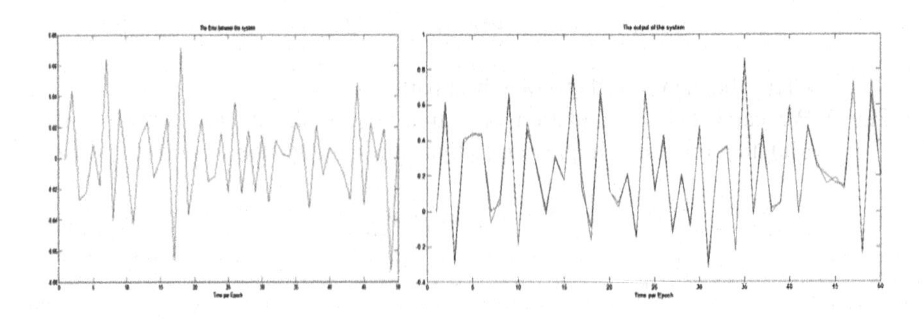

Table 1 Comparison of CPU time used during training operation

Nonlinear system	FLANN-CPSO	FLANN-PSO	FLANN-GA	FLANN-MLP-BP
Trained at −30 dB	0.310	0.320	13.5090	14.6123
Trained at −10 dB	0.320	0.330	12.4080	13.2128

6 Conclusion

A novel Clonal-PSO-based FLANN model has been proposed in this paper for the identification of a particular nonlinear system. Two other FLANN models have also been successfully employed to solve this challenging and interesting problem. The performance of the proposed model has been evaluated and compared, using CPU time. The simulation study reveals that the proposed algorithm has less computational complexity and a faster convergence rate. Further study in this field may include the performance evaluation of the proposed model applied to some other practical systems. The proposed algorithm may also be applied for inverse modeling operations.

References

1. Pao, Y.H.: Adaptive Pattern Recognition and Neural Networks. Addison-Wesley Reading, MA (1989)
2. Sadegh, N.: An orthonormal neural network for function approximation. IEEE Tans. Neural Netw. **4**, 982–988 (1993)
3. Narendra, K.S., Parthasarathy, K.: Identification and control of dynamical systems using neural networks. IEEE Trans. Neural Netw. **1**, 4–26 (1990)
4. Patra, J.C., Kot, A.C.: Nonlinear dynamic system identification using Chebshey functional link artificial neural network. IEEE Trans. Syst. Man Cybern. B: Cybern **32**(4) (2002)
5. Mishra, S.K., Panda, G., Majhi, R.: A comparative performance assessment of a set of multiobjective algorithms for constrained portfolio assets selection, swarm and evolutionary computing. Elsevier **16**, 38–51 (2014)
6. Mishra, S.K., Panda, G., Meher, S., Majhi, R., Singh, M.: Portfolio management assessment by four multiobjective optimization algorithm. IEEE International conference (RAICS-2011), pp. 326–331. Trivandrum, India, 23–25 Sept 2011
7. Eberhart, R., Kennedt, J.: Anew optimizer using particle swarm theory. In: The 6th Int'l Symposimon Micromachine and Human Science, pp. 39–43. Nayoga, Japan (1995)
8. Jiang, J.J., Qin, A.K., Suganthan, P.N., Baskar, S.: Comprehensive learning particle swarm optimizer for global optimization of multimodal functions. IEEE Trans. Evol. Comput. **10**, 281–295 (2006)
9. Lee, T.T., Jeng, J.T.: The Chebyshev polynomial based unified model neural networks for functional approximation. IEEE Trans. Syst. Man Cybern. B **28**, 925–935 (1998)
10. Mishra, S.K., Panda, G., Meher, S.: Chebyshev functional link artificial neural networks for denoising of image corrupted by salt and pepper noise. Int. J. Recent Trends Eng. **1**(1), 413–417 (2009) (Academy publisher, Finland. ISSN: 1797-9617)
11. Mishra, S.K., Panda, G., Meher, S., Sahoo, A.K.: Exponential functional link artificial neural networks for denoising of image corrupted by Gaussian noise. In: IEEE International Conference on Advanced Computer Control (ICACC-2009), pp. 355–359. Singapore, 22–24 Jan 2009

12. Nanda, S.J., Panda, G., Majhi, B.: Improved identification of nonlinear dynamic systems using artificial immune system. In: Annual IEEE India Conference (INDICON), pp. 268–273 (2008)
13. Nanda, S.J., Panda, G., Majhi, B., Tah, P.: Improved identification of nonlinear MIMO plants using new hybrid FLANN-AIS model. In: IEEE International Conference on Advance Computing (IACC), pp. 141–146 (2009)
14. Tan, Y.: Clonal particle swarm optimization and its applications. In: IEEE Congress on Evolutionary Computation, pp. 2303–2309 (2007)

A Steganography Scheme Using SVD Compressed Cover Image

Kshiramani Naik, Shiv Prasad and Arup Kumar Pal

Abstract Steganography is one of the popular tools for secure transmission of confidential data through public channels like the Internet. The common approach of steganography technique is to use any meaningful uncompressed multimedia like image or video as a carrier. The Internet comprises limited channel bandwidth so multimedia data like image or video are considered as a compressed form before their transmission through Internet. So in this paper, we have opted compressed image as a cover media for embedding the secret message and the modified compressed image components will preserve the visual content when it will be reconstructed. Singular value decomposition was employed in cover image for compression. In order to enhance the security, the secret message was embedded into compressed image components where the coefficients were not selected in sequence order. So in the proposed work, a straightforward message extraction process will not be applicable. We have tested the proposed scheme on some standard test images and satisfactory results were achieved.

Keywords Arnold cat map · Least significant bit substitution · Singular value decomposition · Steganography

---

K. Naik (✉) · S. Prasad · A.K. Pal
Department of Computer Science and Engineering, Indian School of Mines, Dhanbad 826004, Jharkhand, India
e-mail: kshiramani@gmail.com

S. Prasad
e-mail: shiv.ismjrf@gmail.com

A.K. Pal
e-mail: arupkrpal@gmail.com

I.K. Sethi (ed.), *Computational Vision and Robotics*, Advances in Intelligent Systems and Computing 332, DOI 10.1007/978-81-322-2196-8_12

1 Introduction

In the digital world, one of the major and serious issues is the preserving secrecy of the important data during communication. Several data protection techniques have been devised to protect the confidentiality of any message. Encryption and steganography are the two popular techniques used for the secure communication of the confidential message. As data encryption techniques provide a stream of meaningless code for transmission that may attract the intruder to recover the message. But steganography method makes unaware about the existence of the confidential message by embedding the secret messages into various multimedia datasets. Already various steganography techniques have been applied for the protection of the confidential data [1–3]. Most of the steganographic techniques have been developed by modifying directly the intensity values of the cover media with tolerable distortion [4–6]. After embedding the secret message into the cover media, it forms stego-object. The general approach in the existing steganographic approaches is hiding the secret information in the uncompressed cover media. But this may degrade the transmission efficiency of the data. Several transformation tools like DCT, SVD, and DWT are well popular and they are used to exploit the presence of compression in multimedia data. Some papers utilize the concept of SVD on the cover media for data hiding purpose [7, 8]. Some SVD-based steganography techniques use singular values of the cover image for secret embedding [9, 10]. The method is robust because it embeds data in low bands of cover in a distributed way but it degrades the image fidelity. In this article, the main objective is to embed the secret message into any compressed cover image. Initially, the cover image is compressed by employing the SVD and subsequently the singular vectors of the cover image matrix are used for hiding the secret image. Before embedding the information into the compressed components of the cover image, the compressed components are shuffled using the generalized ACM which enhance the security mechanism by clubbing the steganography and cryptography.

The rest of this paper is organized as follows. In Sect. 2, some related fundamentals like singular value decomposition (SVD), Arnold cat map (ACM), and least significant bit insertion method (LSB) are defined and discussed. Section 2.3 discusses the proposed Steganography method in detail. The simulation results and security analysis are presented in Sect. 3. Finally, the conclusions are stated in Sect. 4.

2 Preliminaries

2.1 Singular Value Decomposition (SVD)

In linear algebra, the SVD is an important factorization of a real or complex matrix. Let A be a general real (complex) matrix with m rows and n columns, with rank r and $r \leq n \leq m$. Then, the A can be factored into three matrices:

$$A = USV^T \qquad (1)$$

Fig. 1 **a** Original Lena image. **b** Reconstructed image from truncated SVD

where U and V are orthogonal (unitary) and $S = \text{diag}(\sigma 1, \sigma 2, \ldots, \sigma r)$, where σi, $i = 1$ to r are the singular values of the matrix with $r = \min(m, n)$ and satisfying $\sigma 1 \geq \sigma 2 \geq \cdots \geq \sigma r$. SVD is very a popular tool for image compression. Figure 1b shows the reconstructed image with consideration of first 128 eigenvalues, and the computed PSNR value with respect to the original image is around 38.01 dB.

2.2 Arnold Transform (ACM)

ACM is a two-dimensional map that randomizes the positions of the pixels of an image instead of changing their values. ACM for an image matrix of size $N \times N$ is expressed as follows:

$$\begin{bmatrix} x' \\ y' \end{bmatrix} = A \begin{bmatrix} x \\ y \end{bmatrix} \text{ mod } (N) \tag{2}$$

where $A = \begin{bmatrix} 1 & p \\ q & 1+pq \end{bmatrix}$ and p, q are the positive integers. The two parameters p, q and the number of iterations to encode the image. The ACM will be influenced by its initial conditions and that the outputs will appear to be random. In decryption phase, the inverse ACM is applied to get back the original positions of the image. Mathematically, the inverse ACM for a matrix of size $N \times N$ is expressed as follows:

$$\begin{bmatrix} x \\ y \end{bmatrix} = A^{-1} \begin{bmatrix} x' \\ y' \end{bmatrix} \text{ mod } (N) \tag{3}$$

where $A^{-1} = \begin{bmatrix} 1+pq & -p \\ -q & 1 \end{bmatrix}$

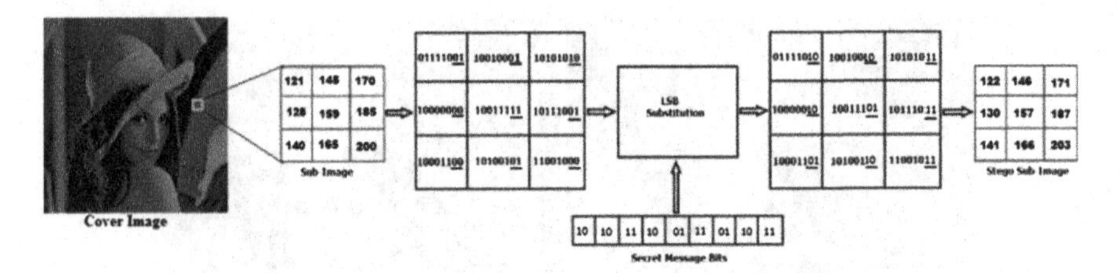

Fig. 2 The LSB substitution scheme

2.3 LSB Substitution Technique

One of the common and the simple steganography approaches to hide the secret message in the digital media is the LSB where a number of the least significant bits of the cover media are replaced by the secret message bits. So, after embedding, human eye cannot find difference between stego-image and cover image. Figure 2 depicts the processes of the LSB substitution method where 18 bits secret message are embedded into the sub-image of size 3 × 3 by replacing first 2 LSB bits of each pixel. After embedding all secret message bits into the cover image, the cover image containing the secret message is termed as the stego-image. In common, up to first 3 LSB bits of cover media are used to hide the secret message bits otherwise the quality of the stego-image will be corrupted greatly. Mathematically, secret message is embedded into the sub-image formula as follows:

$$x' = x - (x \bmod 2^k) + s \tag{4}$$

The secret message can be extracted by the following equation:

$$s = x' \bmod 2^k \tag{5}$$

where x is the pixel value of the cover image, k-bit LSB from secret message s, and x' is the modified stego-image.

2.4 Proposed Scheme

In the proposed scheme, for transmission efficiency only the orthogonal matrices of the cover image matrix are selected as carrier. For high security purpose, the data embedding is also not in sequence order to put difficulty for the intruder during communication.

2.5 Encoding Phase

In this phase, the message bits of the secret image are embedded into the orthogonal matrices of the cover image before transmission. The detail approach is described below (Fig. 3).

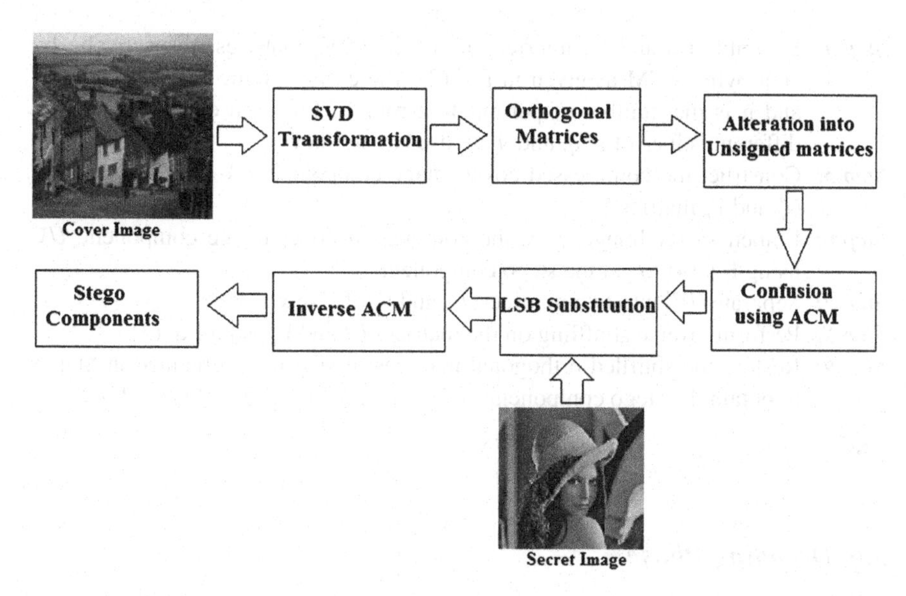

Fig. 3 Block diagram of encoding method

Algorithm: Image_Encoding
Input: A cover image C of size $M \times M$, a secret image S of size $N \times N$
Output: Two stego components U_s and V_s of size $M \times k$ and associated parameters like $\mathrm{Min}(U_{x\times x})$ and $\mathrm{Max}(U_{x\times x})$.

Begin:

Step 1: Apply SVD on the cover image, $C = USV^T$ where $U_{M\times M}$ and $V_{M\times M}$ are orthogonal matrices and S is diagonal matrix
Step 2: Truncate U and V matrices into $U_{M\times k}$ and $V_{M\times k}$ so that the matrices can be reshaped as square matrices like $U_{x\times x}$ and $V_{x\times x}$, respectively.
where k is the number of eigenvectors, and eigenvalues are selected from S, U, V and S matrix, respectively.
Step 3: Perform quantization for suitable representation. The matrices, $U_{x\times x}$ and $V_{x\times x}$, are converted into 8 bit unsigned integer matrices (U_n and V_n) using the following equations, respectively:

$$U_n = \left\lfloor \frac{U_{x\times x} - \mathrm{Min}(U_{x\times x})}{\mathrm{Max}(U_{x\times x}) - \mathrm{Min}(U_{x\times x})} \times 255 \right\rfloor$$

$$V_n = \left\lfloor \frac{V_{x\times x} - \mathrm{Min}(V_{x\times x})}{\mathrm{Max}(V_{x\times x}) - \mathrm{Min}(V_{x\times x})} \times 255 \right\rfloor$$

where $\mathrm{Min}(\bullet)$ and $\mathrm{Max}(\bullet)$ denote the derivation of minimum and maximum value, respectively.

Step 4: Scramble U_n and V_n matrices into U'_n and V'_n matrices, respectively, by employing ACM as given in Eq. (2) where p, q are the ACM parameters and n is the number of permutation rounds. For each encryption round, different values of p, q, and n are used as sub-keys

Step 5: Construct the compressed cover image component, UV by concatenating U'_n and V'_n matrices

Step 6: Embed secret Image, S in the compressed cover image component, UV using Eq. (4) to get the stego component, UV'

Step 7: Separate UV' into two matrices, U'_s and V'_s of size $x \times x$

Step 8: Perform inverse shuffling on the matrices U'_s and V'_s using Eq. (3)

Step 9: Restore the shuffled orthogonal matrices of size $x \times x$ obtained in Step 8 to obtain the stego components, U_s, and V_s, respectively, of size $M \times k$

End

2.6 Decoding Phase

In the decoding phase, the reverse process of the decoding phase is applied on the stego components. The secret message can be extracted using Eq. (3) from the stego components. The stego components and the secret parameters (p, q, and n) used in the encoding phase are given as input in decoding phase.

3 Experimental Results and Analysis

In the proposed scheme, a number of experiments are performed with a gray scale cover image of size 512×512 pixels (Fig. 4a) with grayscale secret images of different sizes. In the proposed technique, three ACM parameters ($p; q; n$) are used as keys. The values so obtained are an indicative measure of the performance and security level attained by the proposed technique. We have tested the perceptibility of the stego-image for the secret image as depicted in Fig. 1a. Figure 4a, e, shows that distortions resulted from embedding are imperceptible to human visual perception. In order to evaluate the image quality for the compressed image and the stego-image, we have computed the peak signal to noise ratio (PSNR). Table 1 shows the PSNR values of different size of secret images. In addition to analyze the statistical similarities of the stego-images with their unmodified version, their corresponding histograms are also given.

Figure 4 shows that the disparity between the constructed histograms and the original histogram is comparatively less. We have obtained high PSNR values for stego-images and also for the secret images. So in our scheme, the PSNR values as well as the visual appearance of the stego-image suggest that the distortion appeared after embedding the secret image into the cover image is reasonably less and imperceptible to human vision.

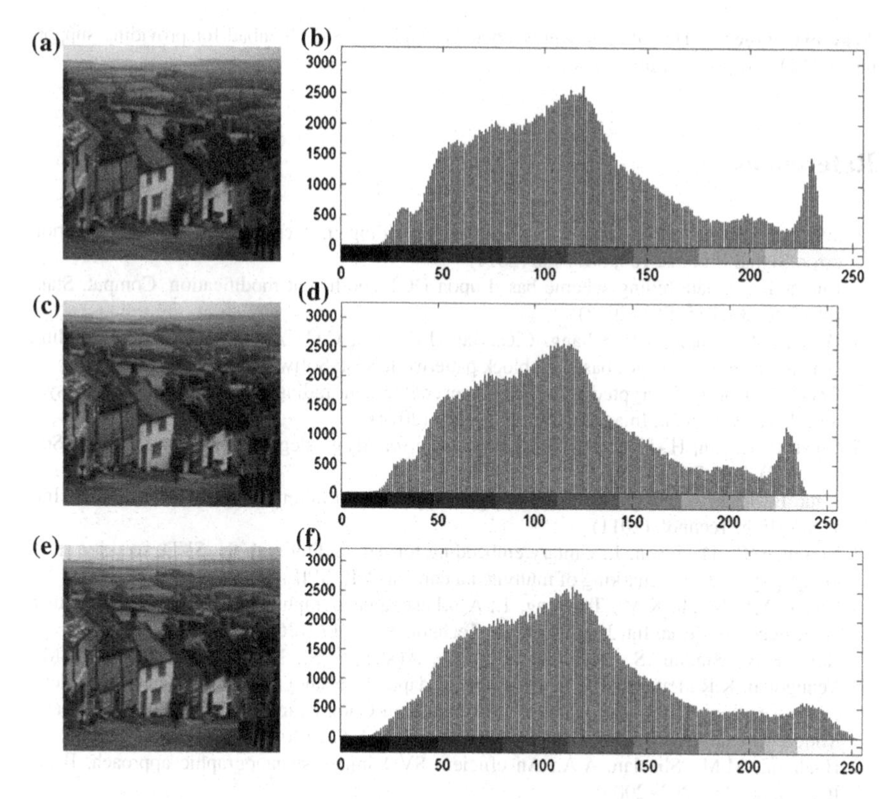

Fig. 4 **a** Cover image. **b** Histogram of cover image. **c** Compressed image. **d** Histogram of compressed image. **e** Stego-image. **f** Histogram of stego-image

Table 1 Calculated values of peak signal to noise ratio

Size of the cover image	Size of the secret image	No. of modified LSB bits	PSNR of the stego-image
512 × 512	342 × 342	2	29.6664
512 × 512	256 × 256	1	32.0955

4 Conclusion

In this paper, a simple LSB substitution method using SVD is proposed. The data hiding takes place only on the orthogonal matrices of the SVD decomposition. The simulation results of the proposed scheme on gray level images have been carried out using MATLAB, and the result analysis shows the effectiveness of the proposed method in terms of PSNR and histogram analysis. In future, the present work can be extended further to improve the hiding capacity of the cover image along with enhancing the security of the secret image.

Acknowledgment The authors express their gratitude to ISM Dhanbad for providing support under TEQIP-II project initiative.

References

1. Wu, X., Sun, W.: High-capacity reversible data hiding in encrypted images by prediction error. Signal Process. **104**, 387–400 (2014)
2. Lin, Y.-K.: A data hiding scheme based upon DCT coefficient modification. Comput. Stan. Interfaces **36**, 855–862 (2014)
3. Wang, C.C., Chang Y.F., Chang, C.C., Jan, J.K., Lin, C.C.: A high capacity data hiding scheme for binary images based on block patterns. J. Syst. Softw. **93**, 152–162 (2014)
4. Chen, S., Horng: Encrypted signal-based reversible data hiding with public key cryptosystem. J. Vis. Commun. Image R. **25**, 1164–1170 (2014)
5. Chen, P.-Y., Lin, H.-J.: A DWT based approach for image steganography. Int. J. Appl. Sci. Eng. **4**(3), 275–290 (2006)
6. Kaur, B., Kaur, A., Singh, J.: Steganographic approach for hiding image in DCT domain. Int. J. Adv. Eng. Technol. (2011)
7. Bergman, C., Davidson, J.: Unitary embedding for data hiding with the SVD, security, steganography, and watermarking of multimedia contents VII. SPIE. 5681 (2005)
8. Chanu, Y.J., Singh, K.M., Tuithung, T.: A robust steganographic method based on singular value decomposition. Int. J. Inf. Comput. Technol. **4**(7), 717–726 (2014)
9. Raja, K.B., Sindhu, S., Mahalakshmi, T.D., Akshatha, S., Nithin, B.K., Sarvajith, M., Venugopal, K.R., Patnaik, I.M.:Robust image adaptive steganography using integer wavelets. In: Proceedings of 3rd International Conference on Communication Systems Software and Middleware and Workshops, COMSWARE'08, pp. 614–621 (2008)
10. Hadhoud, M.M., Shaalan, A.A.: An efficient SVD image steganographic approach. IEEE ICCES, pp. 257–262 (2009)

Particle Swarm Optimization-based Functional Link Artificial Neural Network for Medical Image Denoising

Manish Kumar and Sudhansu Kumar Mishra

Abstract In this paper, a new computationally fast and efficient adaptive digital image filter has been proposed for denoising of digital medical image corrupted with additive white Gaussian noise. A particle swarm optimization-based functional link artificial neural network (FLANN) has been applied for this interesting and challenging problem. The three others competitive networks based on artificial neural network such as multilayer perceptron (MLP), direct linear artificial feed-through neural network (DLFANN), and LMS-based FLANN have also been applied for this purpose. The quantitative analysis of the proposed algorithm has been carried out by taking the peak signal to noise ratio (PSNR) and mean square error (MSE) as two parameters. Experimental results demonstrate the efficiency and effectiveness of the proposed algorithm.

Keywords Medical image · Functional link artificial neural network · Particle swarm optimization · Peak signal to noise ratio

1 Introduction

The medical image may be corrupted during telemedicine due to the noisy communication channel. The presence of noise degrades the medical image quality affecting many of its inherent features that alleviates the problem of disease

Fifth International Conference on Computational Vision and Robotic, 10th August, Bhubaneswar, India.

M. Kumar (✉) · S.K. Mishra
Department of Electrical and Electronic Engineering, Birla Institute of Technology, Mesra, Ranchi, India
e-mail: manish.guptasssss007@gmail.com

S.K. Mishra
e-mail: sudhansu.nit@gmail.com

I.K. Sethi (ed.), *Computational Vision and Robotics*, Advances in Intelligent Systems and Computing 332, DOI 10.1007/978-81-322-2196-8_13

diagnosis, recognition, etc. Hence, to eliminate this problem, medical image should be free from different noises. This important job of medical image denoising can be done by different fixed filters such as mean, median, and rank order mean. However, all these filters are suitable for fixed channel where system parameters are not changing with time. But, in real practice, most of the channels are time varying. To circumvent different limitations of fixed filters, adaptive filters are designed that adapt themselves to the changing conditions.

ANN has proved to be a potential network to perform the tasks in a highly nonlinear environment. MLP is one of the variants of ANN whose weights are trained by applying back propagation (BP) algorithm [1]. In DLFANN structure incorporates, a set of linear terms is introduced in a network for modeling both linear and nonlinear systems simultaneously [2]. The functional link artificial network (FLANN) was proposed by Pao [3]. In the structure of FLANN, hidden layer is absent and whole structure can be represented as a single layer ANN. In this work, it is shown that the FLANN can be used for function pattern classification, quicker rate of convergence, etc. The FLANN model has lesser computational complexity than both MLP and DLFANN networks because of no hidden layer. It is quite capable of forming arbitrarily complex decision regions by generating nonlinear decision boundaries [4]. A FLANN using the Chebyshev and exponential function expansion has been proposed and reported to denoise the image [5, 6]. In this paper, Chebyshev expansion is used for the functional expansion of FLANN, and the particle swarm optimization (PSO) is used for selecting extended inputs and weights. The primary purpose of this paper is to highlight the effectiveness of the proposed PSO-based single layer ANN structure to solve the problem of Gaussian noise cancelation from the medical images. Applications of some other adaptive filters for medical image denoising have also been reported [7–11].

The paper is organized as follows: Different structures of the ANN filter and their implementation for artifact cancelation in image data have been discussed in Sect. 2. Section 3 presents the simulation studies of proposed methods followed by the conclusion in Sect. 4.

2 Structure of ANN Filters

2.1 Multilayer Perceptron

The multilayer perceptron is one of the feed-forward networks, which have one or more hidden layers. All the nodes in all layers (except the input layer) of the MLP contain a nonlinear function. The weighted sum of outputs of a lower layer is passed through the nonlinear function of a node in the upper layer to produce its output. The mean square error (MSE) is used as cost function and the back propagation (BP) algorithm attempts to minimize the cost function by updating all the weights of the network [2] (Fig. 1).

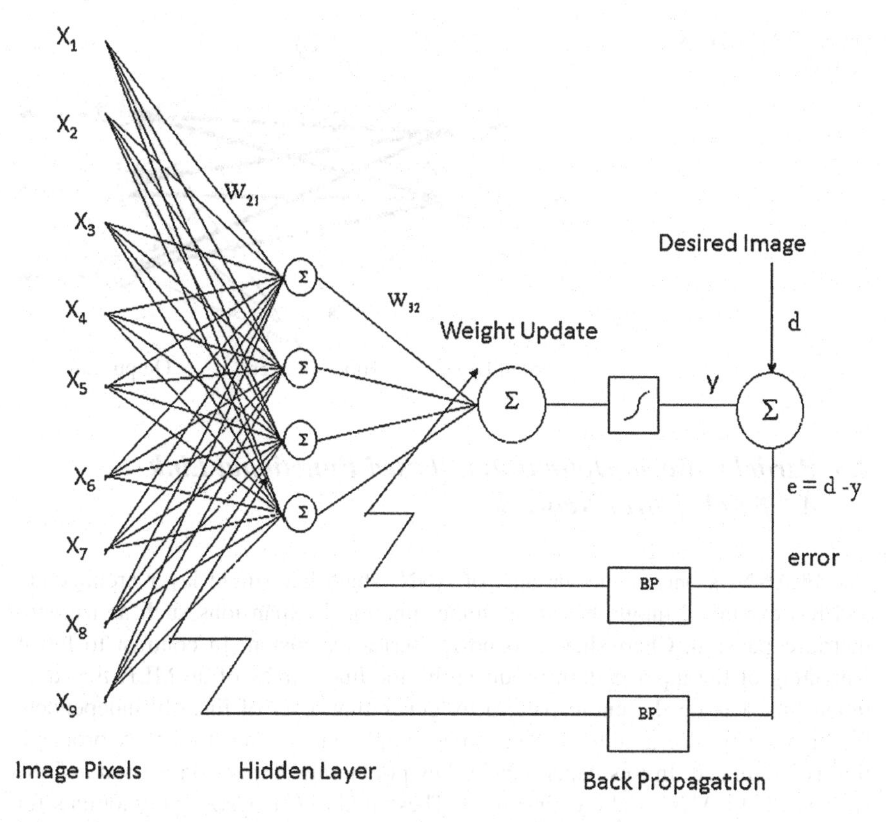

Fig. 1 Multilayer neural network

2.2 Direct Linear Feed-through Neural Network

The process of noise cancelation may have both linear and/or nonlinear behavior over its working range. Hence, it appears more appropriate to have a network structure that is capable of handling both linearity and nonlinearity. The DLFANN configuration which combines conventional NN architecture with a set of linear terms, to produce a network, can handle both linear and nonlinear behavior at the same time [5] (Fig. 2).

In this network, neurons associated with input layer are connected directly with neuron of hidden as well as output layer. The arrangements are shown by thick and thin solid lines, respectively. These direct interconnections between input and output neurons are accountable for handling linear terms. The computational complexity of the network is increased as weights corresponding to direct interconnection between the input and output neurons are added. This increases the complexity of network and the training of network also take more CPU time to update.

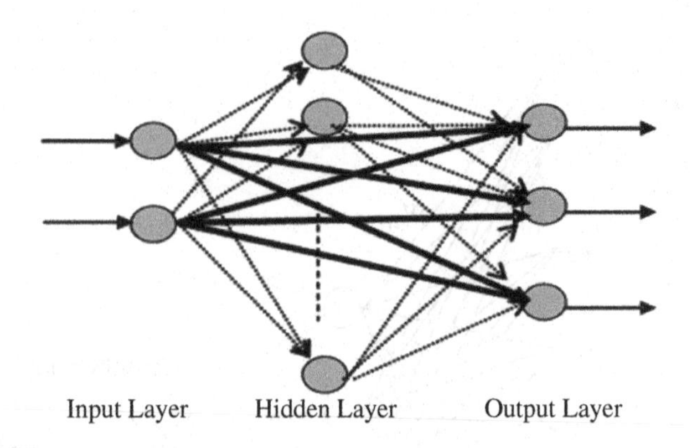

Fig. 2 The DLFANN network

2.3 Particle Swarm Optimization-based Functional Link Artificial Neural Network

The FLANN is one of the variants of ANN which has single layer architecture and has expanded inputs based on some functional expansions such as trigonometric expansion, Chebyshev, and power series expansion. In contrast to linear weighting of the input pattern produced by the linear links of an MLP, the functional link acts on the entire pattern by generating a set of linearly independent functions [12]. The architecture becomes simple and the need of back propagation is eliminated. In this paper, Chebyshev polynomial expansion is used and the structure C-FLANN is shown in Fig. 3. These higher Chebyshev polynomials for $-1 < x < 1$ may be generated using the recursive formula given by

$$T_{n+1} = 2T_n(x) - T_{n-1}(x) \tag{1}$$

The first few Chebyshev polynomials are given by

$$\begin{aligned} T_0(x) &= 1 \\ T_1(x) &= x \\ T_2(x) &= 2x^2 - 1 \\ T_3(x) &= 4x^3 - 3x \\ T_4(x) &= 8x^4 - 8x + 1 \end{aligned} \tag{2}$$

In this paper, some of the expanded inputs of FLANN model are chosen by applying PSO. The weights of the model are chosen by applying PSO. The idea of particle swarm optimization was first introduced by Kennedy and Eberhart in [13]. This algorithm is formulated by observing the social behavior of school of fish or focus of birds. Due to the robustness, simplicity and low computational complexity PSO is applied to many other interesting fields of research [4, 14]. In PSO, each solution is considered by a particle and ith particle (X_{id}), where d is the dimension of the search space. The ith particle of the swarm population has its

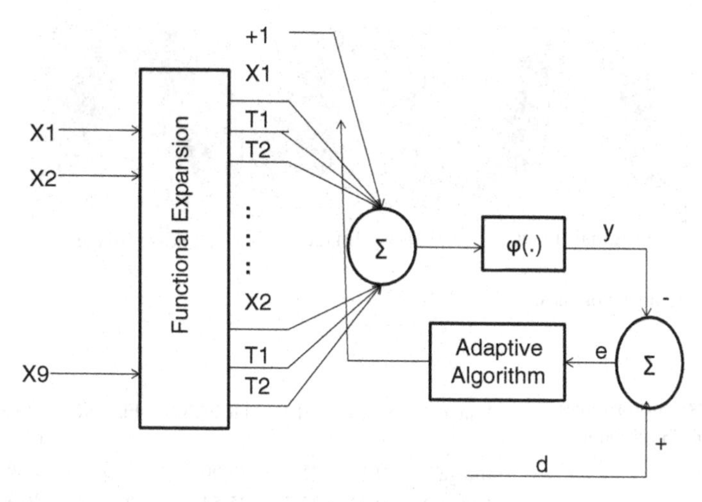

Fig. 3 Chebyshev functional link multilayer perceptron architecture

best position (P_i) that generates the highest fitness value. The global position (P_g) is the best particle that gives best fitness value in the entire population. V_i is the current velocity of ith particle [15]. The position and velocity can be expressed in the next iteration are mathematically expressed as

$$V_i(d) = wV_i(d) + C_i(d) \times \text{rand} \times (P_i(d) - X_i(d)) + C_2 \times \text{rand} \times \left(P_g(d) - X_i(d)\right) \tag{3}$$

$$X_i(d) = X_i(d) + V_i(d) \tag{4}$$

3 Simulation Studies

The MATLAB environment has been implemented for simulation studies. Simulation analysis was carried out with several iterations and to compare performance of various variants of MLPs, DLFANN, and FLANN for denoising of medical image. The proposed method is carried out with X-ray image of chest which degraded by additive white Gaussian noise. Gaussian noise having a continuous energy spectrum of all frequencies and for simulation mean, variance value, i.e., 0 and 0.01 has been selected respectively. In our simulation, we set the structure of MLP, DLFANN as {9-4-1} and FLANN as {9-1}. The order of the polynomial used to obtain expanded inputs is gradually increased so that a minimum number of inputs are used and complexity of the neural network is minimized.

The learning rate of MLP and DLFANN is set to 0.03 and that of FLANN is set to 0.1. The number of iterations was set to be 3,000 for all models. The BP learning algorithm has been used for MLP. In DLFANN, the weight of a direct link between input layer and output layer is trained by LMS algorithm and others

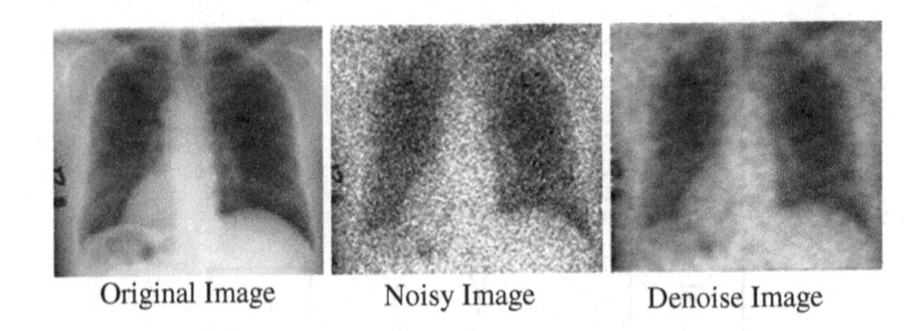

Fig. 4 X-ray image of chest

Table 1 Result for filters in terms of PSNR value

Image	Noisy	MLP	DLFANN	FLANN	PSO-FLANN
Image 1	20.18	25.62	25.66	26.5	26.6
Image 2	20.34	25.74	25.54	26.2	26.3
Image 3	20.13	25.17	25.32	26.1	26.1

are trained by using the BP algorithm. The PSO technique is used for training of FLANN. The training targets were normalized within value [0, 1]. This process continues iteratively until full pattern of the image get processed and whole process continues for 100 times to find MSE as well as PSNR value. The size of X-ray image has been taken 256×256 for the training and testing (Fig. 4).

3.1 Peak Signal to Noise Ratio

In this work, computer simulation is carried out to compare the PSNR value of filtered images obtained from these adaptive models. In this case, the Gaussian noise having mean 0 and variance 0.01 is taken. Let $X(m, n)$, $Y(m, n)$ is the noise-free image, noisy image, respectively. The size of the image of $M \times N$ pixels, where $m = 1, 2, 3 \ldots M$ is the number of rows and $n = 1, 2, 3 \ldots N$ is the number of columns. Then the formula for PSNR and MSE is shown as (Table 1)

$$\mathrm{MSE} = \frac{\sum_{m=1}^{M}\sum_{n=1}^{N}((m,n) - Y(m,n))^2}{\mathrm{M} \times \mathrm{N}} \text{ and } \mathrm{PSNR} = 10.\log_{10}\left(\frac{1}{\mathrm{MSE}}\right) \quad (5)$$

4 Conclusion

The PSO-based FLANN model is proposed for denoising of for medical image corrupted by Gaussian noise. To solve this challenging problem, three other variants of ANN such as MLP, DLFANN, and LMS-based FLANN have also been successfully applied. The comparison of performance of these networks includes evaluation of two performance metrics such as MSE and PSNR. It can be concluded that the proposed PSO-FLANN model reduces the computational complexity with improved performance.

References

1. Haykin, S.: Neural Networks. Maxwell Macmillan, Ottawa (1994)
2. Dash, P.P., Patra, D.: Evolutionary neural network for noise cancellation in image data. Int. J. Comput. Vision Robot. **2**(3), 206–217 (2011)
3. Pao, Y.H.: Adaptive Pattern Recognition and Neural Network. Addison-Wesley Reading, MA (1989)
4. Mishra, S.K., Panda, G., Meher, S., Majhi, R., Singh, M.: Portfolio management assessment by four multiobjective optimization algorithm. In: IEEE International Conference (RAICS-2011), pp. 326–331. Trivandrum, India, 23–25 Sept 2011
5. Mishra, S.K., Panda, G., Meher, S.: Chebyshev functional link artificial neural networks for denoising of image corrupted by salt and pepper noise. Int. J. Recent Trends Eng. **1**(1), 413–417 (2009) (Academy publisher, Finland. ISSN:1797-9617)
6. Mishra, S.K., Panda, G., Meher, S., Sahoo, A.K.: Exponential functional link artificial neural networks for denoising of image corrupted by gaussian noise. In: IEEE Computer Society International Conference on Advanced Computer Control, pp. 355–359 (2009)
7. Muller, H., Michoux, N., Bandon, D.: A review of content-based image retrieval systems in medical applications-clinical benefits and future directions. Int. J. Med. Inform. **73**(1), 1–23 (2004)
8. Sharif, M., Jaffar, M.A., Mahmood, M.T.: Optimal composite morphological supervised filter for denoising using genetic programming: application to medical resonance image. In: Elsevier Conference on Engineering Applications of Artificial Intelligence, pp. 78–89 (2014)
9. Suzuki, K., Horiba, I., Sugie, N.: Efficient approximation of neural filters for removing quantum noise from images. IEEE Trans. Signal Process. **50**(7), 1787–1799 (2002)
10. Jiang, J., Trundle, P., Ren, J.: Medical image analysis with a artificial neural network. Elsevier J. Comput. Med. Imaging Graph. **34**, 617–631 (2010)
11. Mohamed, S.M.R., Jayanthi, R.B.: Speckle noise removal in ultrasound images using particle swarm optimization technique. In: IEEE-International Conference on Recent Trends in Information Technology Chennai, pp. 926–931 (2011)
12. Majhi, B., Sa, P.K.: FLANN based adaptive threshold selection for detection of impulse noise in image. Elsevier Int. J. Electron. Commun. **61**, 478–484 (2007)
13. Kennedy, J., Eberhart, R.: Particle swarm optimization. In: Proceedings of the Fourth IEEE International Conference on Neural Networks, pp. 1942–1948. IEEE Service Centre, Perth, Australia (1995)
14. Mishra, S.K., Panda, G., Manjhi, R.: A comparative performance assessment of a set of multiobjective algorithm for constrained portfolio assets selection. Elsevier Swarm Evol. Comput. **16**, 38–51 (2014)
15. Jansi, S., Subasini, P.: Particle swarm optimisation based total variation filter for image denoising. J. Theor. Appl. Inf. Technol. **57**, 169–173 (2013)

Multi-robot Area Exploration Using Particle Swarm Optimization with the Help of CBDF-based Robot Scattering

Sanjeev Sharma, Chiranjib Sur, Anupam Shukla and Ritu Tiwari

Abstract Robot area exploration is a very important task in robotics because it has many applications in real-life problem. So, this is always a very interesting field for researches. This paper presents a new method for multi-robot area exploration. Here, first the environment is divided into partition. In each partition, the robot is deployed randomly. Each partition is explored separately by robot. For the movement of robots, well-known particle swarm optimization algorithm is used. Here mainly concentrated on the multi-robot-coordinated exploration for unknown search spaces where decisions made by bio-inspired algorithms for movement and thus helping in exploration. For better and fast exploration, robot should be scattered in different directions; for this purpose, new clustering-based distribution method is used. The proposed method is tested on different simulated environments that are considered as indoor and outdoor environments. Different parameters such as move, coverage, energy, and time are calculated. The results show that method works well for both environments.

Keywords Area exploration · Target selection · Particle swarm optimization · Clustering-based distribution factor

S. Sharma (✉) · C. Sur · A. Shukla · R. Tiwari
ABV-Indian Institute of Information Technology and Management, Gwalior, India
e-mail: sanjeev.sharma1868@gmail.com

C. Sur
e-mail: chiranjibsur@gmail.com

A. Shukla
e-mail: dranupamshukla@gmail.com

R. Tiwari
e-mail: tiwariritu2@gmail.com

I.K. Sethi (ed.), *Computational Vision and Robotics*, Advances in Intelligent Systems and Computing 332, DOI 10.1007/978-81-322-2196-8_14

1 Introduction

The problem of area exploration deals with robots to maximize knowledge over a specific area. These applications provide all areas in which the control of intelligent autonomous mobile robot acts as a main role. Ability, sensing, learning, and action are very important factors of such systems, and using such systems is inclining to extensive toward more complicated tasks.

Exploration is the act of moving through an unknown environment building a map that can be used for subsequent navigation [1]. Exploration of unknown environment is an important topic in multi-robot research due to its wide real-world applications, such as search and rescue, hazardous material handling, and military actions [2].

Exploration problem is defined in terms of covering area in such a way to acquire more information about the world as much as less time. The exploration of an area may be done by one robot or a team of robots. When the exploration [3, 4] is performed by a group of robots, then this will certainly explore the environment faster than as compared to single robot, but there are many issues involved when dealing with team of robots. However, a multi-robot system is not n-turned up productive as compared to single robot system due to various kinds of factor [5]. First issue may be space restrictions of the environment forcing the robots to act together so that robots might collide on its way to exploration. Second issue may be the interference among robots and third may be the repeated exploration of the area by many robots due to absence of communication and coordination between the robots.

Many literatures are available related to multi-robot area exploration.

In [6], the new multi-robot coordinated algorithm that applies a global optimized strategy based on k-means clustering is given. Their objective is to guarantee to balanced and sustained exploration of big work space. Here, robot is also assigned in separate area and efficiently reduce the average waiting time for those area. So, these areas waiting time is to be minimum. Their work leads to lowest variance of regional waiting time and the lowest variance of regional exploration percentage.

In [7], particle swarm optimization is used to efficiently chose frontier. So, this work is also used as frontier. Here, the area is divided into subarea. When robot is assigned to the particular subarea, it will not enter the area of other robot. Each robot has two states; first is exploration and second is walking. PSO is used to move the robot.

In [8, 9] also, flocking-based approach is given for the exploration of area.

In [10], robots bid for frontiers, based on the exploration cost and expected information gain. This will cause the communication cost in negotiation grows dramatically when the number of robots increase.

In [11], J. Vazquez and C. Malcolm present a behavior-based architecture for multi-robot exploration and mapping. The architecture is designed to guide the exploration in a decentralized fashion constrained to maintain local short-range communication in a mobile ad hoc network. The behaviors are designed to enhance global performance and are triggered based on local information. The robots are encouraged to move together forming a mobile network and sharing relevant information for the team.

The rest of the paper is organized as follows. Section 2 discusses the methodology for multi-robot area exploration. Section 3 discusses particle swarm optimization. Sections 4 and 5 discuss the directional movement and flow graph, respectively. Parameters' setup is discussed in Sect. 6. Experimental parameters and setup are discussed in Sect. 7. At last, we are concluding the results in Sect. 8.

2 Methodology for Multi-robot Area Exploration

Here, the process of multi-robot area exploration is divided into some phases such as map representation, map partition, deployment of robot, and movement and exploration.

2.1 Environment Representation and Input Map

Here, we are assuming that the environment is represented by grid map. Grid maps discretize the environment into so-called grid cells. Each cell stores information about the area it covers [12]. Each cell in the grid map is divided into three categories—free, occupied, and unknown.

A cell *is free* if it is known to contain no obstacles and is *occupied* if it is known to contain one or more obstacles. All other cells are marked *unknown.* The detection of the three states is done by the different colors of pixels of those belonging to the area obstacle. In the path panning process, the path will be the sequence of free cell. In the area exploration work, objective is to cover all area. So no cell will be unknown. It will be either free or occupied (Fig. 1).

Fig. 1 Environment representation

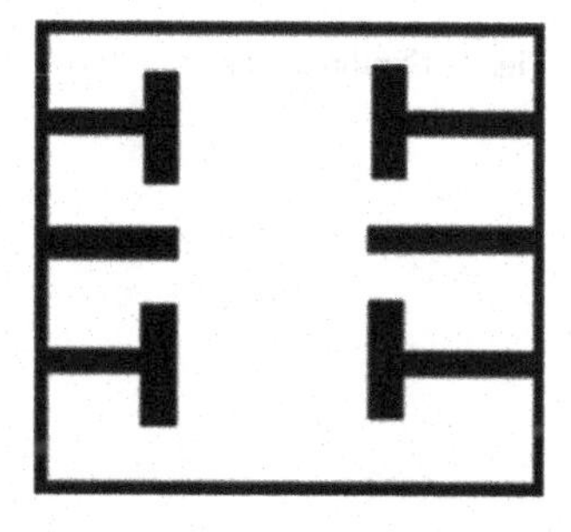

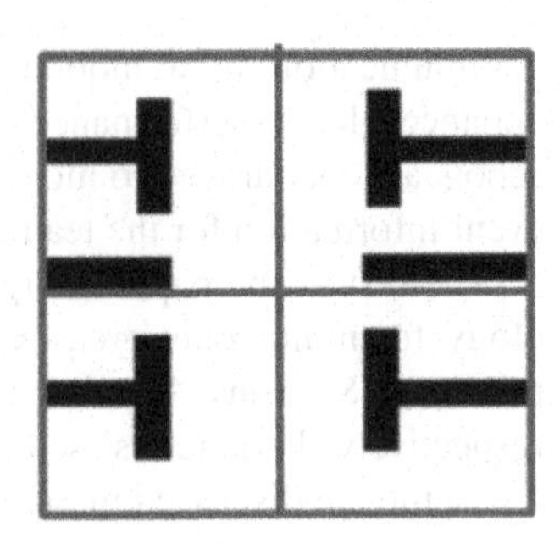

Fig. 2 Map division

2.2 *Map Partition*

Map partition is necessary to avoid the repetition of coverage and for the parallel processing. In this work, map is divided by coordinated bases, and each partition is like square or rectangular bases. Each partition will be treated as small map. So, each map can be explored simultaneously. In Fig. 2, each separate partition is showed.

2.3 *Deployment of Robot in Each Partition*

Robot can be start from any point as a starting point to explore the area. But it is good to start from the corner. In this work, exploration is start from lower left corner from each partition.

2.4 *Movement and the Exploration*

For the movement and the exploration, the well-known nature-inspired algorithm is used. In each partition, the robot is using new clustering-based distribution factor (CBDF) to cover the area.

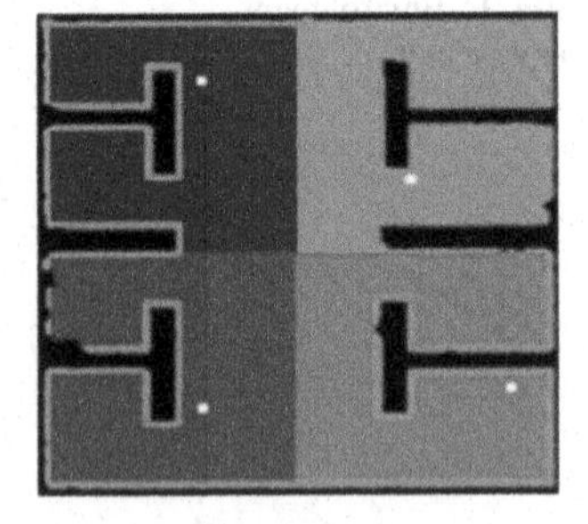

Fig. 3 Explored map

2.5 Output Map After Exploration

Figure 3 shows the results that we are getting after applying proposed multi-robot area exploration method.

3 Modified Particle Swarm Optimization

Particle swarm optimization [13–15] is the population-based global search optimization technique. That is inspired by the flocking of the bird or fish schooling. It is developed by Kennedy and Eberhart in 1995.

PSO can be applied in many optimization problems because it is a simple concept, unique searching mechanism, and computational intelligence. Each particle represents a solution. In each step, particle gets closer to optimum solution by its own behaviors and social behaviors of other particles.

The position and velocity update formula for each particle is given by these equations.

$$x_i(t+1) = x_i(t) + v_i(t+1) \tag{1}$$

$$v_i(t+1) = \omega v_i(t) + c_1\varphi 1(P_{\text{ibest}} - x_i(t)) + c_2\varphi 2(P_{\text{gbest}} - x_i(t)) \tag{2}$$

where $\varphi 1$ and $\varphi 2$ are random variables, that is uniformly distributed within [0, 1]. c_1 and c_2 are the weight change factors. ω is the inertia weight. P_{ibest} represents the best position of the particle, and P_{gbest} is the best position of the swarm.

In multi-robot area exploration, the objective is to cover all area by reaching up to many intermediate targets. Here, this PSO algorithm guides us to move the robot to cover the area.

For exploration task, PSO can be suggest or considered as described below.

A big concern is the global best and the inertia best, and more importantly, they do not exist as the solution is not complete and it is very difficult to determine any temporary fitness regarding it. But, we can actually or up to a certain extent determine the fitness function approximately using the remaining distance for each agent left to reach the target when the target is known. So, less the distance left better is the particle, but in reality, in constrained environment, it is half true and there are much more than it is anticipated.

However in case of target searching, what we did is determination of the area coverage made by an agent and more the area covered more is its fitness. In this way, the positional vector is mapped with the fitness vector of the PSO particle. Rest of the algorithm goes as it is traditionally and as the value of c_1 and c_2 decides the significance amount to be taken from each term, its value need to be determined efficiently.

4 Directional Movement

Clustering-based distribution factor is the new concept which guides the robot in all direction to search the target. In clustering-based distribution factor, the whole region is divided into a number of regions say x, and from any position of the workspace, we can have x direction if we consider that the region head or cluster head is the final point and the position of the agent as the initial point. Using this phenomenon, the robots get the directionality factor and a way to get out of the local region or local optima. There are two kinds of clustering-based distribution factor.

4.1 Scheme 1—Directional Scattering Effect

Directional scattering effect (DSE) is provided to guide the robot to go pass a certain explored area, and when there is requirement of changing a region, come out of a trap or an enclosure, etc. The direction factor is derived out of the cluster head and normally selected randomly to explore new direction and outings.

4.2 Scheme 2—Zig-Zag Search Effect

Zig-Zag search effect (ZSE) is like the directional scattering effect in all respect but mostly used in local search. Here, the direction changes more frequently and

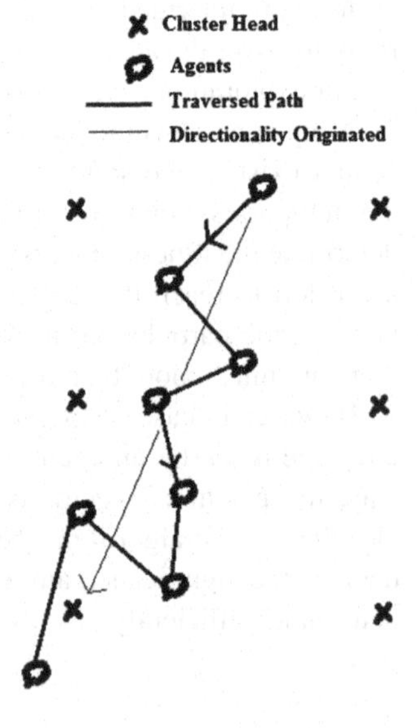

Fig. 4 Directional scattering effect

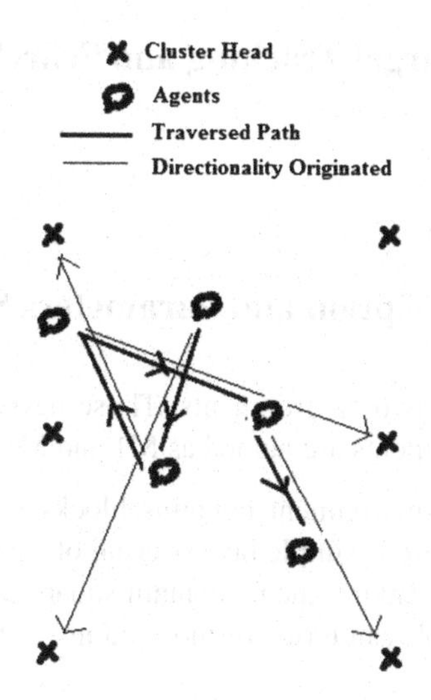

Fig. 5 Zig-zag search effect

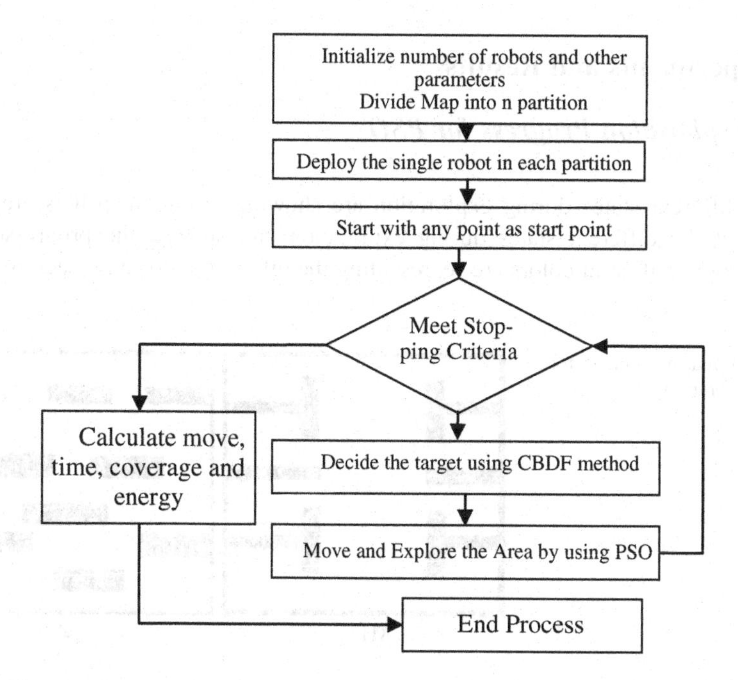

Fig. 6 Flow graph

thus helping more hovering over the workspace. Here, also the cluster heads-based direction is chosen dynamically and randomly.

Figures 4 and 5 show the diagram for directional scattering effect and zig-zag effect.

5 Flow Graph for Target Tracking and Searching

See Fig. 6.

6 Environment Description and Parameters Setup

Here, we are considering two environments. Those have different obstacle structures and density. Environments are named as M1 and M2 as shown in Fig. 7.

M1 Environment—This environment contains blocks that we can consider like indoor. Here, the structure is simple but covering of each block is important.

M2 Environments—Here, environments contain square blocks as obstacles. So, it is relatively more complex in terms of more number of obstacles that are exists (Table 1).

7 Experiments and Results

7.1 Exploration Progress for PSO

Here, different states during exploration are showing. These snapshots are collected at the different states during exploration and showed the progresses in exploration. Different colors are representing the different partitions. Each portion

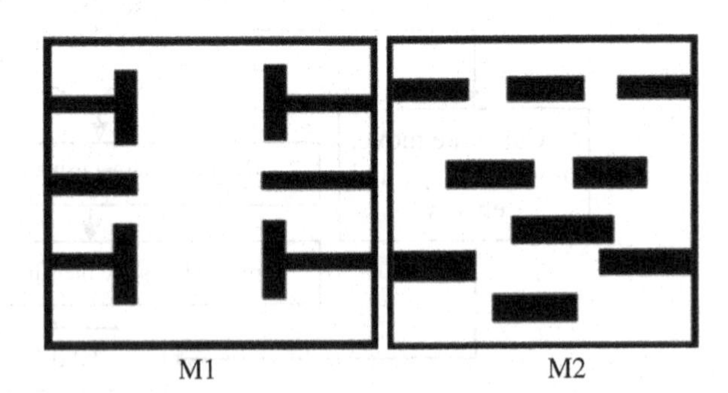

Fig. 7 Environment map for area exploration

Table 1 Parameters for experiments

Parameters	Value
Map size	500 × 500
Team size	4, 6 and 8
Sensor range	2, 4, 6

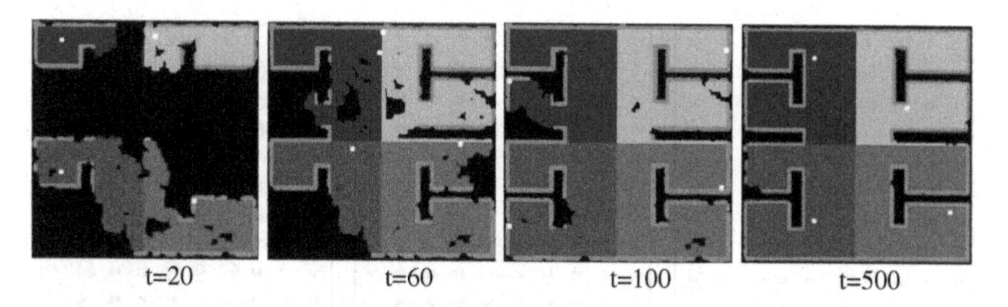

Fig. 8 Exploration map by PSO algorithm for M1

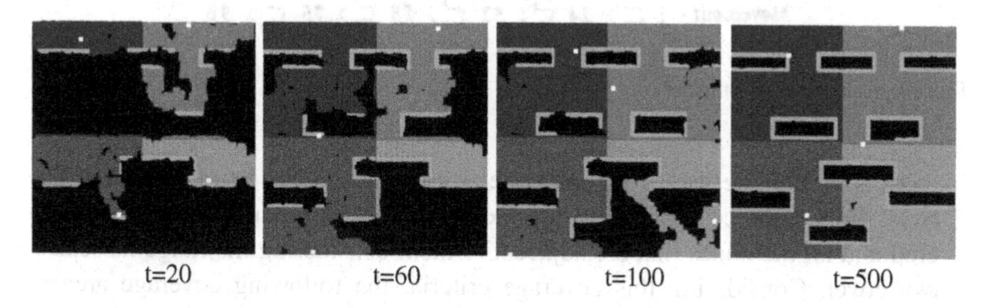

Fig. 9 Exploration map by PSO algorithm for M2

is covered by different robots. Figures 8 and 9 show the progress of exploration by using PSO for environment M1 and M2. Here, number of robots to explore the area is 4.

7.2 Parameters Calculation During Exploration

Here, comparisons are shown between different algorithms. Different parameters such as moves, time, coverage, and energy are taken.

Moves Calculation—Let us consider that an agent moves through the following nodes 1 → 24 → 52 → 58 → 75 → 90 as shown in Fig. 10. In this case, if the agent has covered 5 nodes, the number of moves counted as (nodes = nodes + 5).

Energy Calculation—Similarly for energy, we consider the number of unit shifts in any direction (only horizontal and vertical). Hence, while covering from node 1 to 24, horizontally there is movement of 3 units and vertically 2 units. This makes a total count of 5 energy consumption. Similarly, for these 5 moves, we have a total energy consumption of 27 energy units.

Coverage Calculation—Coverage is an important aspect of exploration for the agents and covers all the nodes it can visualize from the nodes it has covered. The coverage area depends on the visibility or sensing range of the agents. Also

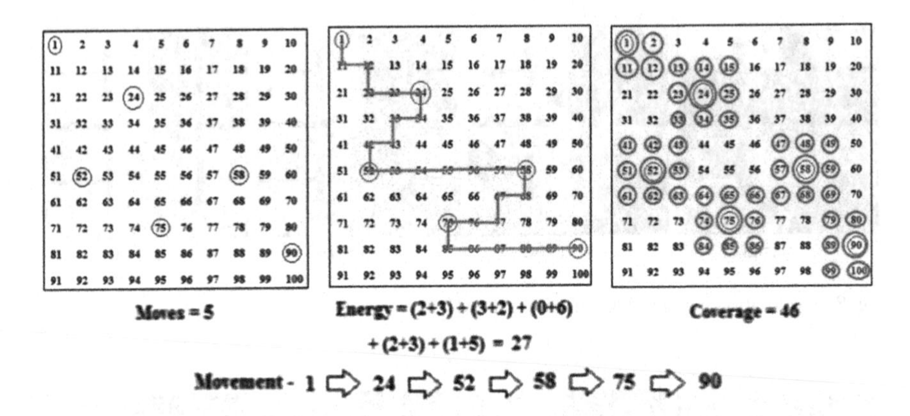

Fig. 10 Calculated parameters

there is no definite way from one node to another and can be of many alternatives. Hence, the coverage area considered of only those area from it has covered and all the nodes that are adjacent to them considering the range its sensor can cover. Considering this coverage criteria, the following coverage area is considered in figure.

Time Calculation—The timing factor, considered here, is the simulation time which various when the number of agents are varied. So, the exact time calculation will be approximately the average timing factor that is the noted time calculation divided by the number of agents

Tables 2, 3, and 4 show the calculated values for 4, 6, and 8 robots, respectively, for different detection ranges. These are calculated for Map1. DR represents the detection range of robots.

Table 2 Calculated values when the number of robots is 4

DR	Coverage	Time	Energy	Move
2	28,726	0.094432	19,470	2,520
4	43,113	0.096487	17,583	2,067
6	63,894	0.105043	17,227	1,880

Table 3 Calculated values when the number of robots is 6

DR	Coverage	Time	Energy	Move
2	40,746	0.0853428	28,150	3,677
4	57,886	0.091617	25,851	3,056
6	76,168	0.0934588	25,037	2,747

Table 4 Calculated values when the number of robots is 8

DR	Coverage	Time	Energy	Move
2	53,505	0.0812386	37,830	4,987
4	70,646	0.088774	32,997	3,932
6	75,810	0.093227	32,068	3,597

8 Conclusion

This paper presents a new method of multi-robot area exploration. Here, the next location for the movement of robot is process by PSO. For scattering of robots in different directions, a new concept is doing that is very effective to cover the area. Four parameters such as move, time, coverage, and energy are calculated. The results show that this method is able to explore the area very well.

References

1. Yamauchi, B.: A frontier-based approach for autonomous exploration. In: IEEE International Symposium on Computational Intelligence in Robotics and Automation, CIRA, pp. 146–151 (1997)
2. Wu, H., Tian, G., Huang, B.: Multi-robot collaboration exploration based on immune network model. In: IEEE/ASME International Conference on Advanced Intelligent Mechatronics, AIM, pp. 1207–1212 (2008)
3. De Hoog, J., Cameron, S., Visser, A.: Role-based autonomous multi-robot exploration. In: Future Computing, Service Computation, Cognitive, Adaptive, Content, Patterns, 2009. COMPUTATIONWORLD'09. Computation World, pp. 482–487 (2009)
4. Marjovi, A., Nunes, J.G., Marques, L., de Almeida, A.: Multi-robot exploration and fire searching. In: IEEE/RSJ International Conference on Intelligent Robots and Systems, IROS, pp. 1929–1934 (2009)
5. Thrun, S., Bücken, A.: Integrating grid-based and topological maps for mobile robot navigation. In: Proceedings of the National Conference on Artificial Intelligence, pp. 944–951 (1996)
6. Puig, D., Garcia, M.A., Wu, L.: A new global optimization strategy for coordinated multi-robot exploration: development and comparative evaluation. Robot. Auton. Syst. **59**(9), 635–653 (2011)
7. Wang, Y., Liang, A., Guan, H.: Frontier-based multi-robot map exploration using particle swarm optimization. In: IEEE Symposium on Swarm Intelligence, pp. 1–6 (2011)
8. Cheng, K., Wang, Y., Dasgupta, P.: Distributed area coverage using robot flocks. In: Nature and Biologically Inspired Computing, pp. 678–683 (2009)
9. Parker, L.E.: On the design of behavior-based multi-robot teams. Adv. Robot. **10**(6), 547–578 (1995)
10. Simmons, R., Apfelbaum, D., Burgard, W., Fox, Moors, M., Thrun, S., Younes, H.: Coordination for multi-robot exploration and mapping. In: Proceedings of the AAAI National Conference on Artificial Intelligence (AAAI), pp. 852–858 (2000)
11. Vazquez, J., Malcolm, C.: Distributed multirobot exploration maintaining a mobile network. Int. Conf. Intell. Syst. **3**, 113–118 (2004)
12. Elfes, A.: Using occupancy grids for mobile robot perception and navigation. Computer **22**(6), 46–57 (1989)
13. Eberhart, R.C., Kennedy, J.: A new optimizer using particle swarm theory. In: Proceedings of the sixth international symposium on micro machine and human science, vol. 1, pp. 39–43 (1995)
14. Pugh, J., Martinoli, A.: Inspiring and modeling multi-robot search with particle swarm optimization. In: Proceeding in Swarm Intelligence IEEE Symposium, SIS, pp. 332–339 (2007)
15. Zhang, Y., Gong, D.W., Zhang, J.H.: Robot path planning in uncertain environment using multi-objective particle swarm optimization. Elsevier J. Neurocomput. **103**, 172–185 (2013)

R–D Optimization of EED-based Compressed Image Using Linear Regression

Bibhuprasad Mohanty, Madhusmita Sahoo and Sudipta Mohapatra

Abstract Image compression technique minimizes the size in bytes of a graphics file without degrading the quality of the image to an unacceptable visual level. The performance is driven by the trade-off between rate (R), distortion (D), and complexity of the coding algorithm. This piece of work deals with the simple yet effective method of image-dependent R–D optimization by optimizing the cost function of the constrained Lagrangian multiplier. A simple linear regression is employed to estimate the cost and the slope of the convex haul in the R–D plane. The image is compressed by SPIHT algorithm after it was subjected to anisotropic diffusion to provide a powerful zero tree pyramid. The Perona–Malik diffusion preserves the edge and boundary information by smoothening the whole image. Extensive simulation for the proposed diffusion platform provides better results in terms of PSNR and visual quality, and the estimation approximates the actual values with minimum residuals.

Keywords Anisotropicdiffusion · SPHIT · EED · R–D optimization · Linear regression · OLS

1 Introduction

The storage of the rapidly increased multimedia products by using less hardware space and transmitting them with lowest possible bandwidth is achieved with reduction in redundancies from the data, called data compression. Image as a source of information and multimedia applications plays the vital and significant

B. Mohanty (✉) · M. Sahoo
Faculty of Engineering, Department of ECE, ITER, Siksha O Anusandhana University, Bhubaneswar, India
e-mail: bibhumohanty@soauniversity.ac.in

S. Mohapatra
Department of E&ECE, Indian Institute of Technology Kharagpur, Kharagpur, India

I.K. Sethi (ed.), *Computational Vision and Robotics*, Advances in Intelligent Systems and Computing 332, DOI 10.1007/978-81-322-2196-8_15

roles. By reducing the redundancies and irrelevancies from the image, the consumption of memory space and bandwidth is reduced. Image compression addresses the problem of reducing the amount of data required to represent the digital image. Image compression demands an effective coding technique to meet the trade-off between bandwidth and storage capacity with the quality of reconstructed image. In natural images, most scenes are relatively blurred in the areas between edges, while large discontinuities occur at edge locations. Thus, the information between edges may be redundant and may be subjected to increased compression. However, the human visual system is highly sensitive to edges and hence an edge-preserving-based image compression technique can produce intelligible images at high compression ratios. By introducing more redundancies in an image by smoothening it by the application of partial differential equation (PDE), better compression can be achieved [1, 2]. But, when we try to smooth the image, it is equally likely that information pertaining features such as edges, line may be lost. Since the human visual system (HVS) is more sensitive to edges, boundary, etc., we propose an edge-enhancing diffusion (EED)-based technique to preserve such features and at the same time to introduce more redundancies [3]. The subsequent experimental results are perceptually impressive and show that an edge detector based on nonlinear PDE gives more stable edges across the scale [4]. Embedded image coding such as SPIHT improves coding performance and also allows a bit stream to be truncated at any point and still decode a reasonably good image [5]. However, due to this arbitrary truncation, the reconstructed image does not necessarily produce optimal image quality. Hence, there is a trade-off between rate and distortion and a need remains for an embedded coder that optimizes rate–distortion at many different truncation points in an encoded bit stream [6]. Rate–distortion optimization (RDO) refers to the optimization of the amount of distortion (D, loss of quality) against the amount of data required to encode, the rate (R). R–D can be described in terms of Lagrangian multipliers. It can also be described by the principle of equal slopes, which states that the coding parameters should be selected so that the rate of change of distortion with respect to bit rate is the same for all parts of the system [7]. In this piece of work, we have tried to substantiate that the EED-based compression not only augers well with time constraints, but also provides a reasonable well and fast R–D estimation.

The rest of this paper is organized as follows. In Sect. 2, we describe the EED-based compression advantages as compared to other diffusion method with reference to our previous work. In Sect. 3, the proposed linear regression-based optimization methodology has been detailed. We have concluded this paper in Sect. 4 with a brief outline of work in future direction.

2 EED-based Image Compression with SPIHT Algorithm

In our previous work [8], we have discussed the EED method and have shown that the wavelet-based transform coding algorithm such as SPIHT achieved superior compression under the diffusion platform. The objective of that work was to show

Table 1 Lena image and corresponding error comparisons

Criteria	Homogenous diffusion	Bi-harmonic diffusion	Tri-harmonic diffusion	EED
MSE	138.7331	137.9309	137.0078	35.0189
AAE	12.1542	12.5261	12.7646	4.2887
SSIM	0.9710	0.9589	0.9491	0.9876
Diffused image				

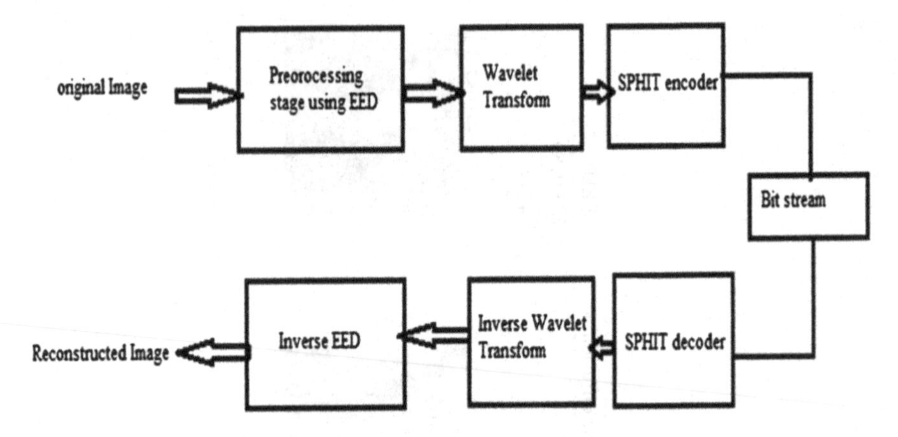

Fig. 1 Proposed SPIHT-based image codec [8]

the preservation of edge information and to provide better PSNR performance. In Table 1, the error performance comparison has been made for Lena image to show that EED has an edge over other diffusion methods [9].

The following is the system overview (Fig. 1) to carry out the proposed image compression scheme. The EED scheme is a preprocessing scheme before the encoder in the transmitter side. By preprocessing the original image, we have smoothened the whole image but the edge. This process of smoothening introduces more redundancy. On transforming the image to the frequency domain, the energy is compacted, and hence, fewer numbers of bits are required to code the

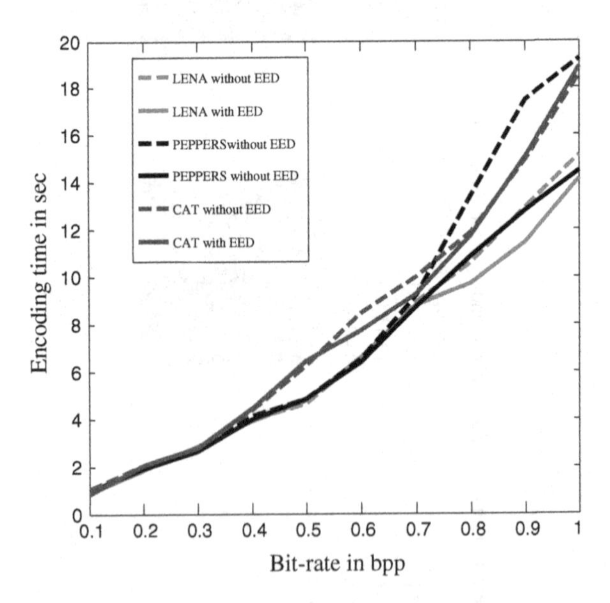

Fig. 2 Encoding time comparison for SPIHT

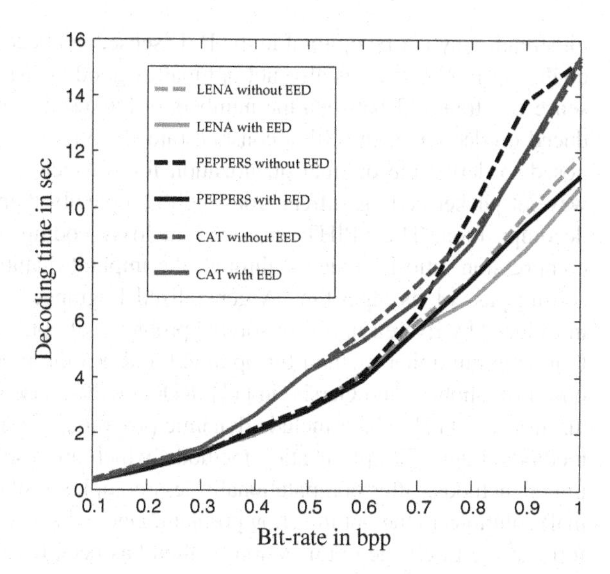

Fig. 3 Decoding time comparison for SPIHT

original one. On the decoder side, the reverse operation takes place. The output of the decoded image is once again subjected to the method of inverse operation of EED to approximate the lost data.

The simulation is carried out using Pentium 2, Core i5 processor with 4-GB RAM and 2.60-GHz clock frequency. Throughout the experiment, MATLAB 2011 has been used by taking different images. The application of wavelet transform may increase the speed of the codec and facilitate its hardware implementation. It is because of the fact that the preprocessed image is a smooth version of the original one except at the point of edges and hence bit plane coding scheme using wavelet transform will require less time for encoding as well as decoding. Figure 2 shows the time taken by the SPIHT codec at different bit rate and the effect of EED on it for three test images under consideration. Figure 3 also provides the decoding time complexity comparison for EED-based codec without only prejudice we are showing that EED-based codec is faster in encoding as well as decoding.

3 Linear Regression for Estimation of RDO with Lagrangian Multiplier

The quality of the reconstructed image or image sequences (videos) can be improved by optimizing the amount of distortion (D) against the rate at which the codec encodes/decodes the image. Rate–distortion optimization (RDO) algorithm judiciously improves the quality of reconstruction, thereby affecting the file size. Embedded image coding allows the bit stream to be truncated at any point arbitrarily or at the point of meeting bit budget. But such truncation points of the encoded bit stream may not be optimal in the R–D sense, and hence, the reconstructed image quality in PSNR sense is also not optimal. A good decision in compression in R–D sense is a trade-off between the numbers of bits used to encode with the error produced on decoding. In [10], a constant rate–distortion slope criterion has been proposed to derive the optimal quantization for wavelet packet coding. However, the wavelet packet coding suffers from high complexity than the standard dyadic tree decomposition. The SPIHT encoder is a lossy coding technique and the desired compression ratio is achieved through the implicit quantization process during the sorting pass of its algorithm. A generalized Lagrange multiplier method was first introduced by Everett in [6] for solving problems of optimal resource allocation. The Lagrange multiplier method for optimal bit allocation in a coding scheme was proposed by Shoham and Gersho in [11]. RDO methods used in video compression are discussed in [12], which include dynamic programming and Lagrange optimization methods. Lagrange optimization methods, which are also known as Lagrange multiplier methods, offer computationally less complex (although sometimes sub-optimal) solutions to the optimization problem. Due to its less complex nature, a specific form of the Lagrange optimization method has been used in rate–distortion optimization of H.264/AVC and H.263 reference software encoders [13, 14].

3.1 Linear Regression

In statistics, linear regression is an approach for modeling the relationship between a scalar dependent variably and one or more explanatory variables denoted X. In linear regression, data are modeled using linear predictor functions, and unknown model parameters are estimated from the data. Such models are called *linear models*. Most commonly, linear regression refers to a model in which the conditional mean of y given the value of X is an affine function of X [15]. Ordinary least-squares (OLS) regression is a generalized linear modeling technique that may be used to model a single response variable which has been recorded on at least an interval scale. The technique may be applied to single or multiple explanatory variables and also categorical explanatory variables that have been appropriately coded.

3.2 Estimation of RDO for EED-based SPIHT Compressed Image

Edge-preserving smoothing and super-resolution are classic and important problems in computational photography and related fields. In this chapter, we intend to formulate the estimation of the EED-based compressed image based on linear regression. In the following, we briefly review the ordinary linear regression (OLS), and subsequently, we present some of our estimated result for the chosen test images. The constraint optimization problem for Lagrangian multiplier can be thought of as a linear relationship.

$$\text{We know, } J = D + \lambda R \tag{3.1}$$

$$\text{Rearranging, } D = -\lambda R + J. \tag{3.2}$$

The relationship between D and R may be represented by using a line of best fit. The applicability of this model stems from the fact that the distortion D and the iteration obey a linear relationship in the processing stage. On the other hand, for the SPIHT encoder, the MSE (D) decreases (conversely, PSNR increases) with the rate. However, the diffusion method also affected the D irrespective of R. Hence, it is pertinent to develop the R–D curve so that the distortion got minimum for a target bit rate. Hence, we propose to apply OLS to predict the distortion D, with the value of λ (which is also called the regression coefficient) to fit the model. We have implemented the linear regression algorithm for three different images (namely Lena, Peppers, and Cat) while computing its error table (Table 2).

In the following table, we tabulate the value of the J and λ for two conditions, with applying the diffusion and without applying the diffusion (Table 3).

The estimation curve and actual curve for 3 different images (Lena,peppers, and cat) in Fig. 4 justify our assumption that EED-based compression has an edge over the other part. The slope and cost is relatively less in case of EED-based compression as compared to compression without EED, and hence the residuals is also comparatively less. The predicted R–D line approaches more toward the minima of the R–D convex null and hence the estimation is better.

Table 2 Error comparison for proposed SPIHT codec

	MSE for Lena		MSE for Peppers		MSE for Cat	
Bitrate	Without EED	With EED	Without EED	With EED	Without EED	With EED
0.1	76.0432	67.1165	88.7468	80.2635	105.4741	92.2795
0.2	38.7401	32.0304	44.1406	36.0938	69.7889	59.8324
0.3	25.6906	20.0547	29.7591	22.0096	50.9930	43.2623
0.4	18.3907	12.9921	23.3903	15.5425	41.1885	34.1238
0.5	14.2461	9.6869	18.5773	10.9434	33.4011	27.4287
0.6	11.8985	7.4296	16.4772	8.9693	27.4132	21.8786
0.7	10.1900	5.8628	14.7777	7.1926	22.6034	18.1776
0.8	8.7595	4.4240	13.5058	5.6957	19.3854	15.4871
0.9	7.5633	3.5759	12.3034	4.4529	16.5941	13.1068
1	6.7242	2.9417	10.7597	3.7028	14.1167	10.8439

Table 3 Estimation of J and λ using OLS

Images	Lena		Peppers		Cat	
	Without EED	With EED	Without EED	With EED	Without EED	With EED
Variables						
J	53.6150	45.8965	61.7087	53.4574	87.2412	75.3842
λ	−57.8008	−53.2456	−62.6635	−61.7652	−85.7188	−75.8948

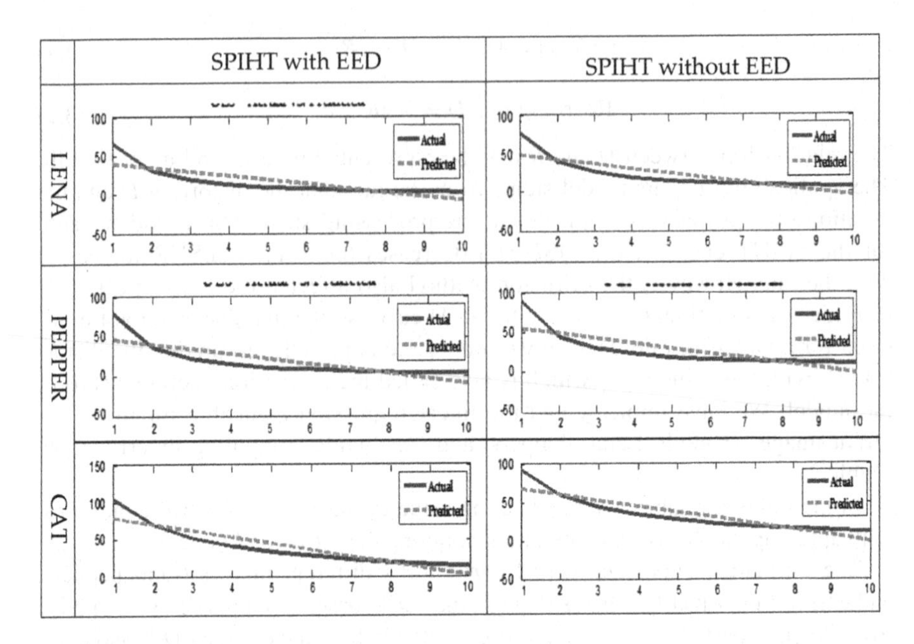

Fig. 4 The actual and estimated R–D curves for test images

4 Conclusion

In this work, the PDE-based nonlinear anisotropic diffusion method, termed as EED, has attracted our attention for its edge-enhancing property. When it is applied as preprocessing tool in a compression domain, it does only the intraregion smoothing, and hence, the edges and the lines in the images are preserved. And the redundancy increases due to the smoothing; it provides better representation of quantized coefficient in the transformed domain. By increasing more number of redundancies, the SPIHT encoder requires less number of bits to produce encoded data stream. Thus, the encoder requires less time for encoding and similarly the decoder also requires less decoding time. This approach can suitably be extended for video coding at low bit rate.

References

1. Sapiro, G.: From active contours to anisotropic diffusion: relations between basic PDEs in image processing. In: Proceedings of ICIP. Lausanne, Switzerland (1996)
2. Perona, P., Malik, J.: Scale-space and edge detection using anisotropic diffusion. In: Proceedings of IEEE Computer Society Workshop on Computer Vision, pp. 16–22 (1987)
3. Shah, J.: A common framework for curve evolution, segmentation, and anisotropic diffusion. In: Proceedings of CVPR, pp. 136–142. San Francisco, CA (1996)

4. Perona, P., Malik, J.: Scale-space and edge detection using anisotropic diffusion. IEEE Trans. Pattern Anal. Mach. Intell. **12**(7), 629–639 (1990)
5. Said, A., Pearlman, W.A.: A new, fast and efficient image codec based on set partitioning in hierarchical trees. IEEE Trans. Circ. Syst. Video Technol. **6**, 243–250 (1996)
6. Everett, H.: Generalized Lagrange multiplier method for solving problems of optimal allocation of resources. Oper. Res. **11**, 399–417 (1963)
7. Ortega, A., Ramchandran, K.: Rate-distortion methods for image and video compression. IEEE Signal Process. Mag. **15**(6), 23–50 (1998)
8. Behera, L., Sahoo, M., Mohanty, B.: Image compression using edge enhancing diffusion. In: Proceedings in Advances in Intelligent Systems and Computing, pp. 457–464. Springer Book Series, Berlin (2015). doi:10.1007/978_81_322_2009_1
9. Guo, Z., Sun, J., Zhang, D., Wu, B.: Adaptive Perona–Malik model based on the variable exponent for image denoising. IEEE Trans. Image Process. **21**(3), (2012)
10. Xiong, Z., Ramchandran, K.: Wavelet packet-based image coding using joint space-frequency quantization. In: IEEE International Conference on Image Processing. Austin, 13–16 Nov 1994
11. Shoham, Y., Gersho, A.: Efficient bit allocation for an arbitrary set of quantizers. IEEE Trans. Acoust. Speech Signal Process. **36**, 1445–1453 (1988)
12. Ortega, A., Ramchandran, K.: Rate-distortion methods for image and video compression. IEEE Signal Process. Mag. **15**(6), 23–50 (1998)
13. Joint Video Team (JVT) of ISO/IEC MPEG and ITU T VCEG. JM7.3 test model CODEC. http://iphome.hhi.de/suehring/tml/
14. ITU-T Study Group 16. Q15-D-65d1, video codec test model, near term, version 10 (TMN10) Draft 1. Finland (1998)
15. Hawkins, D.H.: On the investigation of alternative regressions by principal component analysis. J. R. Stat. Soc. Ser. C **22**(3), 275–286 (1973)

Comparative Analysis of Image Deblurring with Radial Basis Functions in Artificial Neural Network

Abhisek Paul, Paritosh Bhattacharya, Prantik Biswas and Santi Prasad Maity

Abstract Radial basis functions are used in many fields of mathematics and image analysis. In this paper, we have used linear RBF, cubic RBF, multi-quadratic RBF, inverse multi-quadratic RBF and Gaussian RBF for the reconstruction of blurred images. Simulations and mathematical comparisons show that Gaussian RBF gives better result with respect to the other RBF methods for images reconstruction in artificial neural network.

Keywords Neural network · Linear RBF · Cubic RBF · Gaussian RBF · Multi-quadratic RBF · Inverse multi-quadratic RBF · Radial basis function

1 Introduction

Different radial basis functions (RBFs) of artificial neural network (ANN) are utilized in various areas such as pattern recognition, optimization and approximation process. Image reconstructions are also being done in many areas in recent days. We have chosen few radial basis function neural networks such as linear, cubic, multi-quadratic, inverse multi-quadratic and Gaussian RBFs for calculation. [1–3]. Some colour images are taken as example and their Y colour components are being introduced for the analysis. The images are then converted into blurred image and taken as for reconstruction with several radial basis function methods.

A. Paul (✉) · P. Bhattacharya · P. Biswas
Department of Computer Science and Engineering, National Institute of Technology Agartala, Agartala, India
e-mail: abhisekpaul13@gmail.com

S.P. Maity
Department of Information Technology, Bengal Engineering and Science University, Shibpur, India

I.K. Sethi (ed.), *Computational Vision and Robotics*, Advances in Intelligent Systems and Computing 332, DOI 10.1007/978-81-322-2196-8_16

Mean-squared error, mean absolute error and peak signal-to-noise ratio are being shown and compared with different hidden layers of radial basis function neural network. In Sect. 2, architecture of RBF neural network is described with various types of radial basis functions. Section 3 shows the computational analysis of radial basis functions with pseudo-inverse technique. In Sect. 4, simulations and result are given, and finally in Sect. 5, we have given the conclusion.

2 Architecture of Radial Basis Function Neural Network

As shown in Fig. 1, RBF neural network architecture consists of three layers: the input layer, hidden layer and output layer. In Fig. 1, inputs are $X = \{x_1,\ldots, x_d\}$ which enter into input layer. Radial centres and width are $C = \{c_1,\ldots,c_n\}^T$ and σ_i, respectively. In hidden layer, $\phi = \{\phi_1,\ldots, \phi_n\}$ are the radial basis functions. Centres are of $n \times 1$ dimension when the number of input is n. The desired output is given by y which is calculated by proper selection of weights. $w = \{w_{11},\ldots, w_{1n},\ldots, w_{m1},\ldots, w_{mn}\}$ is the weight. Here, w_j is the weight of i-th centre [1, 2].

$$y = \sum_{i=1}^{m} w_i \phi_i \tag{1}$$

$$\phi(x) = \|x - c\| = r \tag{2}$$

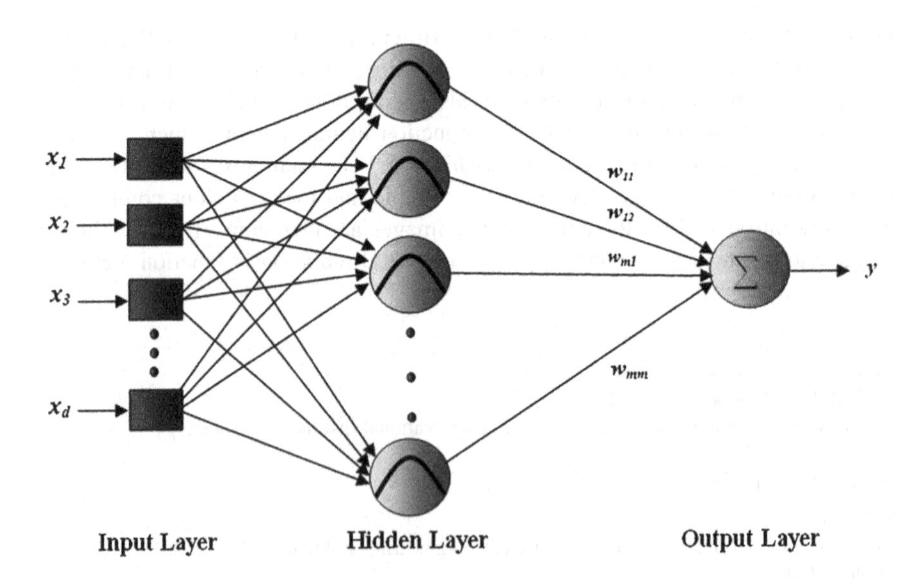

Fig. 1 Radial basis function neural network architecture

Radial basis functions such as linear, cubic, thin plane spline and Gaussian are given in the in the Eqs. (2), (3), (4) and (5), respectively.

- Linear RBF:

$$\phi(x) = r \tag{3}$$

- Cubic RBF:

$$\phi(x) = r^3 \tag{4}$$

- Multi-quadratic:

$$\phi(x) = (r^2 + \sigma^2)^{1/2} \tag{5}$$

- Inverse multi-quadratic:

$$\phi(x) = (r^2 + \sigma^2)^{-1/2} \tag{6}$$

- Gaussian:

$$\phi(r) = \exp(-\frac{r^2}{2\sigma^2}) \tag{7}$$

Radial basis functions such as linear, cubic, multi-quadratic, inverse multi-quadratic and Gaussian are given in the in the Eqs. (3), (4), (5), (6) and (7), respectively.

3 Computational Analysis

We have taken some colour images. After that, we extracted the red, green and blue colour components from the colour images. Then, we calculated Y colour component [4] from the Eq. (9). Relation between Y components and R, G and B colour component is given in Eq. (8). As we have applied linear, cubic, multi-quadratic, inverse multi-quadratic and Gaussian RBF in artificial neural network, RBF needs optimal selections weight and centres. We have calculated all the method with pseudo-inverse technique [5]. As input, we have chosen peppers and jet plane images. Matrix sizes of all the input images are of 256 × 256 pixels for every analysis and experiment.

$$\begin{bmatrix} Y \\ I \\ Q \end{bmatrix} = \begin{bmatrix} 0.299 & 0.587 & 0.144 \\ 0.596 & -0.274 & -0.322 \\ 0.211 & -0.523 & 0.312 \end{bmatrix} \begin{bmatrix} R \\ G \\ B \end{bmatrix} \tag{8}$$

$$Y = 0.299R + 0.587G + 0.114B \tag{9}$$

4 Simulation and Result

Colour images with Y colour components are taken for the simulation. In this paper, we have experimented through two different colour images which are peppers and jet plane, respectively. We have simulated mean-squared error (MSE), mean absolute error and peak signal-to-noise ratio (PSNR) of these images with linear, cubic, multi-quadratic, Inverse multi-quadratic and Gaussian RBFs. We have used MATLAB 7.6.0 software [6] for the analysis and simulation process.

In Fig. 2, original peppers image, Y colour component of peppers image, blurred image and desired output are shown. In Fig. 3, original jet plane image, Y colour component of peppers image, blurred image and desired output are shown.

In Fig. 4, comparison of MSE of peppers image with linear RBF, cubic RBF, multi-quadratic RBF, inverse multi-quadratic RBF and Gaussian RBF is shown when number of hidden layer is set to 80. In Fig. 5, comparison of MAE of peppers image with linear RBF, cubic RBF, multi-quadratic RBF, inverse multi-quadratic RBF and Gaussian RBF is shown when number of hidden layer is set to 80. In Fig. 6, comparison of PSNR in dB of peppers image with linear RBF, cubic RBF, multi-quadratic RBF, inverse multi-quadratic RBF and Gaussian RBF is shown when number of hidden layer is set to 80.

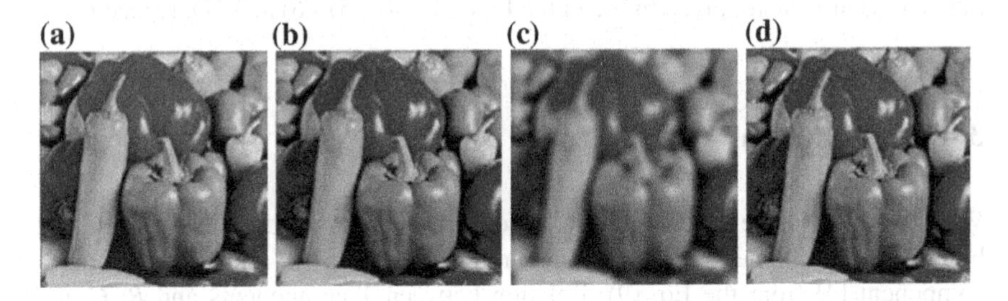

Fig. 2 Peppers images. **a** Original colour Image. **b** Y colour component. **c** Blurred image. **d** Desired output

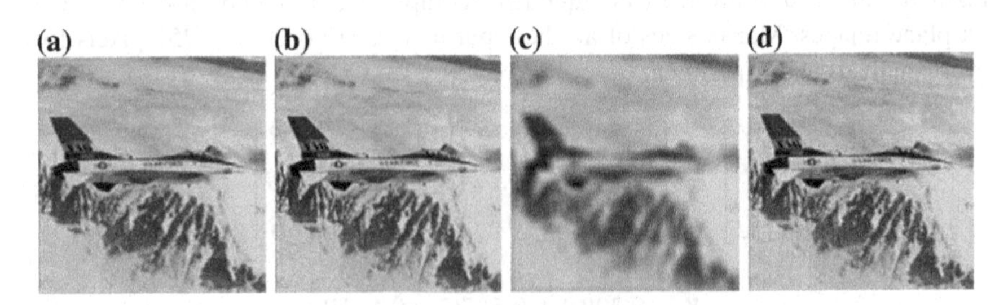

Fig. 3 Jet plane images. **a** Original colour image. **b** Y colour component. **c** Blurred image. **d** Desired output

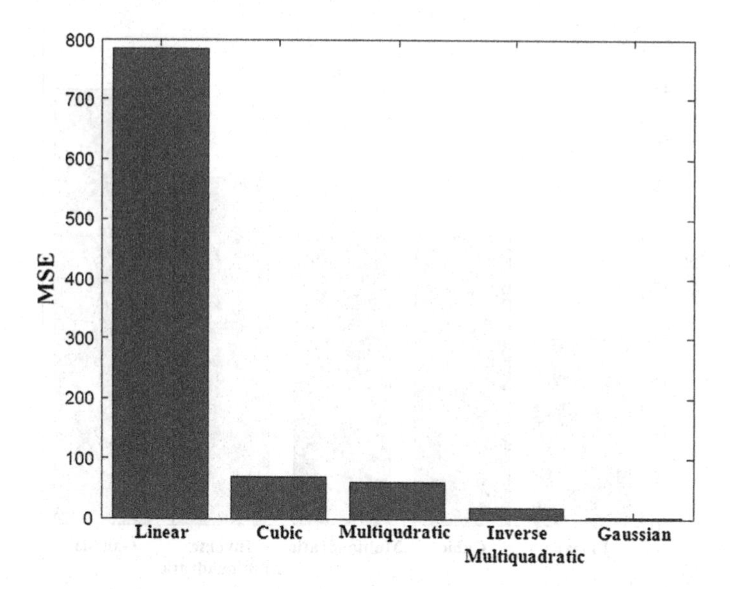

Fig. 4 Comparison of MSE of peppers image with linear RBF, cubic RBF, multi-quadratic RBF, inverse multi-quadratic RBF and Gaussian RBF when number of hidden layer is 80

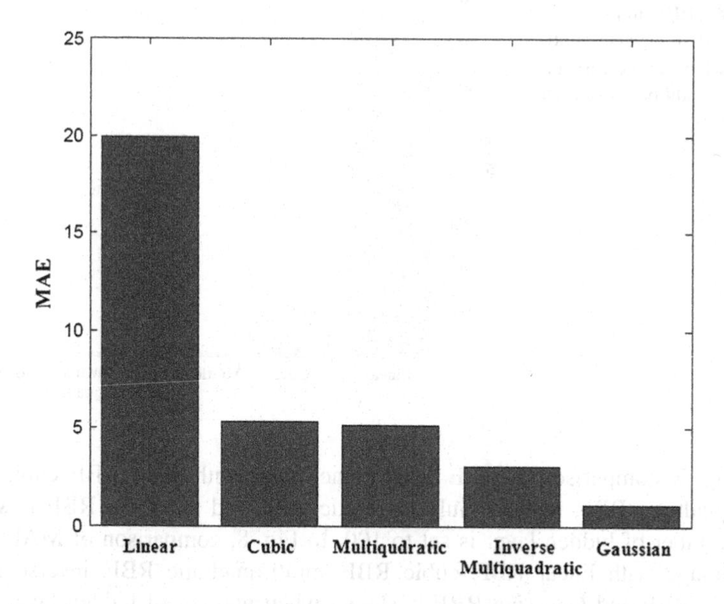

Fig. 5 Comparison of MAE of peppers image with linear RBF, cubic RBF, multi-quadratic RBF, inverse multi-quadratic RBF and Gaussian RBF when number of hidden layer is 80

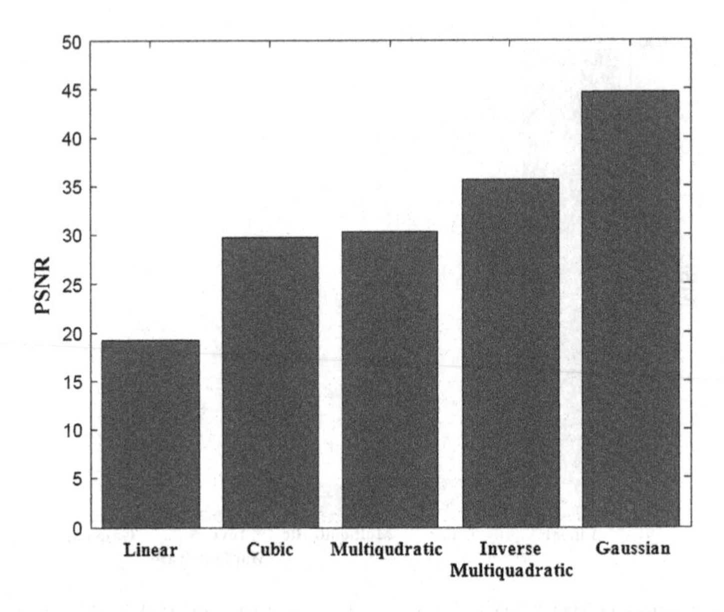

Fig. 6 Comparison of PSNR (dB) of peppers image with linear RBF, cubic RBF, multi-quadratic RBF, inverse multi-quadratic RBF and Gaussian RBF when number of hidden layer is 80

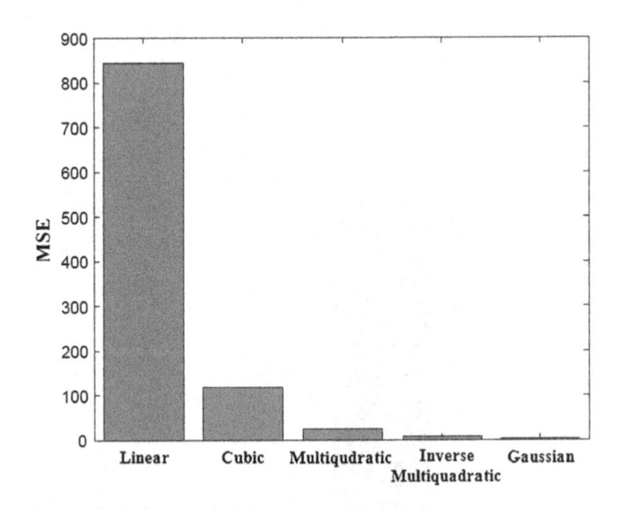

Fig. 7 Comparison of MSE of jet plane image with linear RBF, cubic RBF, multi-quadratic RBF, inverse multi-quadratic RBF and Gaussian RBF when number of hidden layer is 100

In Fig. 7, comparison of MSE of jet plane image with linear RBF, cubic RBF, multi-quadratic RBF, inverse multi-quadratic RBF and Gaussian RBF is shown when number of hidden layer is set to 100. In Fig. 8, comparison of MAE of jet plane image with linear RBF, cubic RBF, multi-quadratic RBF, inverse multi-quadratic RBF and Gaussian RBF is shown when number of hidden layer is set to 100. In Fig. 9, comparison of PSNR in dB of jet plane image with linear RBF, cubic RBF, multi-quadratic RBF, inverse multi-quadratic RBF and Gaussian RBF

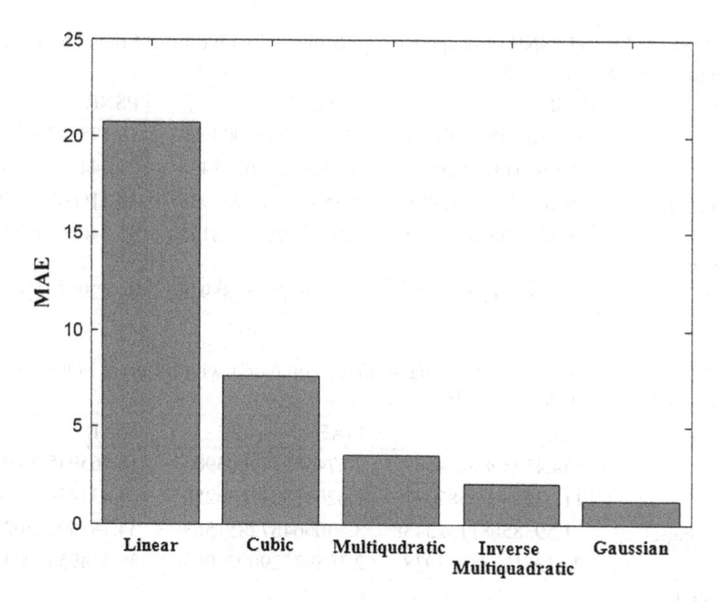

Fig. 8 Comparison of MAE of jet plane image with linear RBF, cubic RBF, multi-quadratic RBF, inverse multi-quadratic RBF and Gaussian RBF when number of hidden layer is 100

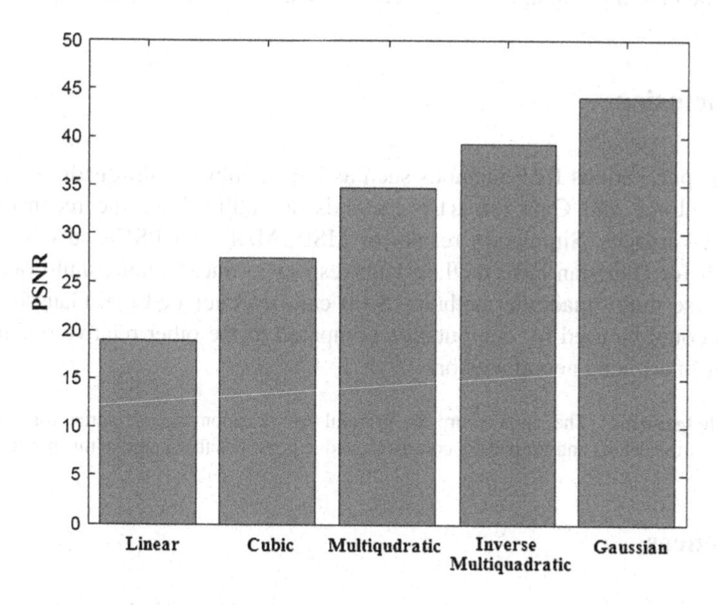

Fig. 9 Comparison of PSNR(dB) of jet plane image with linear RBF, cubic RBF, multi-quadratic RBF, inverse multi-quadratic RBF and Gaussian RBF when number of hidden layer is 100

Table 1 MSE, MAE and PSNR of peppers image compared with different radial basis functions when number of hidden layer is 80

RBF function	MSE	MAE	PSNR
Linear	785.6081085205078	19.906753540039063	19.178744035692365
Cubic	68.966049194335938	5.322372436523438	29.744450132644154
Multi-quadratic	60.461593627929688	5.153842163085938	30.316007708760154
Inverse multi-quadratic	18.157836914062500	3.062530517578125	35.540162497356611
Gaussian	2.239364624023438	1.075912475585938	44.629555476333849

Table 2 MSE, MAE and PSNR of jet plane image compared with different radial basis functions when number of hidden layer is 100

RBF function	MSE	MAE	PSNR
Linear	844.4535980224609	20.746597290039063	18.865045704025110
Cubic	117.3908996582031	7.620178222656250	27.434459298968051
Multi-quadratic	21.593856811523438	3.540664672851563	34.787501436207144
Inverse multi-quadratic	7.613510131835938	2.105972290039063	39.314954308488012
Gaussian	2.563232421875000	1.241607666015625	44.042923731146892

is shown when number of hidden layer is set to 100. In Tables 1 and 2, simulated results of MSE, MAE and PSNR (dB) with different hidden layers of RBF are given when the input images are peppers image and jet plane image respectively.

5 Conclusion

In this paper, various RBF methods such as linear, cubic, multi-quadratic, inverse multi-quadratic and Gaussian RBF methods are utilized for the reconstruction of blurred images. Simulation results of MSE, MAE and PSNR (dB) give better result for Gaussian RBF method with respect to linear, cubic, multi-quadratic and inverse multi-quadratic methods. So it can be concluded that Gaussian RBF method could be used for computation compared to the other relative radial basis methods in artificial neural network.

Acknowledgements The authors are so grateful to the anonymous referee for a careful checking of the details and for helpful comments and suggestions that improve this paper.

References

1. Mao, K.Z., Huang, G.-B.: Neuron selection for RBF neural network classifier based on data structure preserving criterion. IEEE Trans. Neural Netw. **16**(6), 1531–1540 (2005)
2. Schölkopf, B., Sung, K.-K., Burges, C.J.C., Girosi, F., Niyogi, P., Poggio, T., Vapnik, V.: Comparing support vector machines with Gaussian kernels to radial basis function classifiers. IEEE Trans. Signal Process. **45**, 2758–2765 (1997)

3. Luo, F.L., Li, Y.D.: Real-time computation of the eigenvector corresponding to the smallest eigen value of a positive definite matrix. IEEE Trans. Circ. Syst. **41**, 550–553 (1994)
4. Liu, Z., Liu, C.: Fusion of the complementary discrete cosine features in the *YIQ* color space for face recognition. Comput. Vision Image Underst. **111**(3), 249–262 (2008)
5. Klein, C.A., Huang, C.H.: Review of pseudo-inverse control for use with kinematically redundant manipulators. IEEE Trans. Syst. Man Cybern. **13**(3), 245–250 (1983)
6. Math Works. MATLAB 7.6.0 (R2008a) (2008)

A Hybrid Approach for Iris Recognition in Unconstrained Environment

Navjot Kaur and Mamta Juneja

Abstract Iris recognition is one of the emerging areas as the demand for security in social and personal areas is increasing day by day. The most challenging step in the process of iris recognition is accurate iris localization. As it significantly affect the further processing of feature extraction and template matching stages. Traditional algorithms work efficiently and accurately in localizing iris from eye images taken in constrained environmental conditions. But their accuracy gets affected when eye images are taken in unconstrained environmental conditions. The proposed algorithm starts with determining the exact location of iris even in the presence of specular highlights and non-ideal environmental conditions. It also works well for noisy iris images, and the accuracy of segmentation stage has increased when intuitionistic fuzzy is applied on the UBIRIS v2 iris images.

Keywords Intuitionistic fuzzy · Fuzzy c-mean · Iris recognition and biometric

1 Introduction

In this technological era, the demand for security in every field is in danger, and it is very challenging to deal with this security-related problem. Biometric technologies can be used to tackle this security problem. These biometric technologies include figure print, palm vein, voice, iris recognition, and many more [1, 2]. There is a large interest in improved, authentic, secure and genuine identification methods which are more useful. The demand in the area of iris is increasing as it

N. Kaur (✉) · M. Juneja
CSE Department UIET, Panjab University, Chandigarh, India
e-mail: navjotkaur2611@gmail.com

M. Juneja
e-mail: mamtajuneja@pu.ac.in

I.K. Sethi (ed.), *Computational Vision and Robotics*, Advances in Intelligent Systems and Computing 332, DOI 10.1007/978-81-322-2196-8_17

is the most unique feature of human body which remains stable throughout life even though the person ages. Iris has the highest degrees of freedom among all the biometric features [1–3]. Some desirable properties such as uniqueness and stability make iris recognition suitable for human identification [1, 3]. Iris as a recognition system was introduced in 1987 for the first time [1]. Many researchers such as Daugman [2–6] and Bowyer [7] proposed many powerful iris recognition algorithms. Some of these algorithms need user cooperation when image is being taken to get high-quality image. These traditional algorithms work efficiently with eye images taken under ideal conditions only. But for using this technology in real-life scenario, it should work effectively and efficiently under non-ideal conditions also [8–10].

The inspiration behind this paper is to propose an algorithm to handle eye images which are noisy and taken under non-ideal situations.

2 Proposed Scheme

In the proposed algorithm, preprocessing of the eye images is done first to remove extra noise and blurring. Steps involved in the proposed scheme are given and shown in Fig. 1.

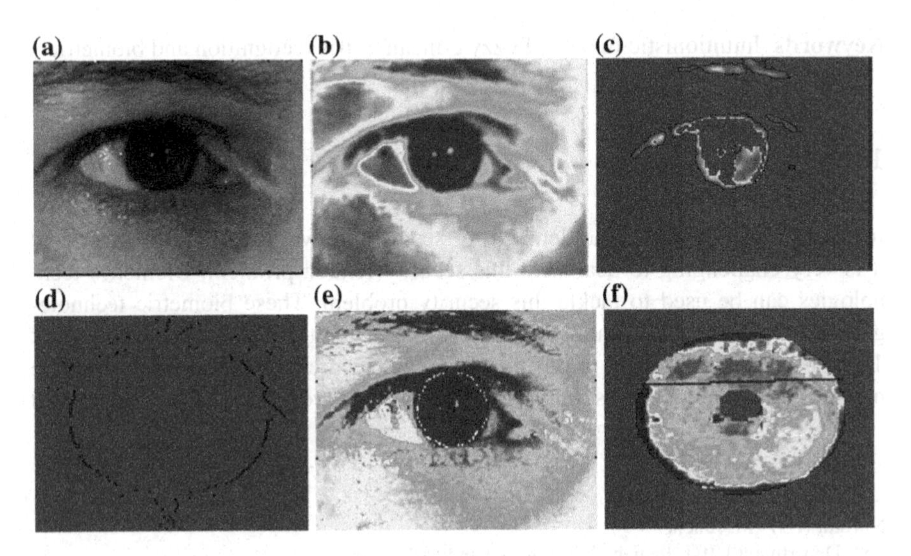

Fig. 1 The stages of proposed algorithm **a** original eye image **b** clustered image after intuitionistic fuzzy c-mean, darkest region represents iris region **c** create image edge map using Canny edge detector **d** apply Canny edge detector to the iris region to get rough iris boundaries **e** apply circular Hough transform to determine iris **f** localization and removal of eyelid and pupil region

2.1 Image Segmentation

Image segmentation is done by using fuzzy c-mean clustering technique [11]. In the concept of Fuzzy set [12], degree of membership is given. However for real-life scenarios, there always exists a degree of hesitation or uncertainty associated with this degree of membership which is not given in any fuzzy set. Thus, the concept of intuitionistic fuzzy set (IFS) was given to introduce the degree of hesitation [13], which is given as B

$$B = \{(x, \mu_B(x), \nu_B(x)) | x \in X\} \quad (1)$$

where $\mu_B(x)$ is degree of belongingness and $\nu_B(x)$ is degree of non-belongingness.

2.2 Edge Detection

Image is handled morphologically to delete small noise present in the image. Canny edge detector is used to find edges [14]. The threshold values for Canny edge detector are adjusted experimentally so that the algorithm works suitably for noisy iris images.

2.3 Circular Hough Transform (CHT)

Circular hough transform (CHT) is applied after the Canny edge detector to get circular iris boundaries. After applying CHT, binary edge image is formed. As eye images of UBIRIS v2 dataset [15] are taken in non-ideal conditions so many noisy factors exists that needs to be removed to avoid errors.

2.4 Eyelid, Eyelash, and Pupil Removal

The lower eyelids are removed by the use of linear Hough transform as most of the occlusions in the lower part of the eye are linear and can be easily removed. Upper eyelids are removed by using Canny edge detection and then applying Hough transform in vertical direction. A line is drawn from the left side of the eye above the pupil area and meeting the other end of the eye, where there are maximum eyelash occlusions. This line is drawn after applying Canny edge detector and getting the exact eyelash edges which are then removed with this line.

Pupil is removed at the end by using CHT to create its boundary.

Table 1 Comparison between the accuracy of the proposed algorithm and the previous algorithms

Method	Accuracy (%)	Time (s)
Daugman	95.22	2.73
Wildes	98.68	1.95
Camus and Wildes	96.78	3.12
S.A. Sahmoud	98.76	1.49
Proposed	98.80	1.36

2.5 Template Matching

The final stage involves comparison of binary iris codes to determine whether they match or not for iris recognition. Comparisons were calculated accurately by Hamming distance (HD), a dissimilarity measure.

3 Experimental Results

The implementation was done on the eye images taken from UBIRIS v2 eye image dataset [15]. The dataset contained images taken under unconstrained conditions. The accuracy and the average segmentation time for the proposed algorithm and various previous algorithms are given in Table 1.

4 Conclusion

This paper proposed a segmentation scheme for noisy eye images based on intuitionistic fuzzy c-mean to segment the eye image into its constituent parts and makes iris localization easy. Experimental results show the accuracy of the proposed scheme in comparatively less time.

References

1. Flom, L., Safir, A.: Iris recognition system. U.S. Patent 4641394, (1987)
2. Daugman, J.: Biometric personal identification system based on iris analysis. U.S. Patent 5291560, (1994)
3. Daugman, J.: High confidence visual recognition of persons by a test of statistical independence. IEEE Trans. Pattern Anal. Mach. Intell. **15**(11), 1148–1161 (1993)
4. Daugman, J.: New methods in iris recognition. IEEE Trans. Syst. Man Cybern. Part B: Cybern. **37**(5), 1167–1175 (2007)
5. Daugman, J.: Statistical richness of visual phase information: update on recognizing persons by their iris patterns. Int. J. Comput. Vis. **45**(1), 25–38 (2001)

6. Daugman, J.: Demodulation by complex-valued wavelets for stochastic pattern recognition. Int. J. Wavelets Multiresolut. Inf. Process. **1**(1), 1–17 (2003)
7. Bowyer, K.W., Hollingsworth, K., Flynn, P.J.: Pupil dilation degrades iris biometric performance. Comput. Vis. Image Underst. **113**(1), 150–157 (2009)
8. Kaur, N., Juneja, M.: A review on iris recognition. In: Recent Advances in Engineering and Computational Sciences (RAECS), pp. 1–5 (2014)
9. Sahmoud, S.A., Abuhaiba, I.S.: Efficient iris segmentation method in unconstrained environment. Pattern Recognit. **46**(12), 3174–3185 (2013)
10. Wildes, R.: Iris recognition: an emerging biometric technology. Proc. IEEE **85**(9), 1348–1363 (1997)
11. Pal, N.R., Bezdek, J.C.: On cluster validity for the fuzzy c-means model. IEEE Trans. Fuzzy Syst. **3**(3), 370–379 (1995)
12. Zadeh, L.A.: Fuzzy sets. Inf. Control **8**, 338–353 (1965)
13. Deschrijver, G., Kerre, E.E.: On the representation of intuitionistic fuzzy t-norms and t-conorms. IEEE Trans. Fuzzy Syst. **12**(1), 45–61 (2004)
14. Canny, J.: A computational approach to edge detection. IEEE Trans. Pattern Anal. Mach. Intell. **8**(6), 679–698 (1986)
15. Proença, H., Filipe, S., Santos, R., Oliveira, J., Alexandre, L.A.: The UBIRIS.v2: a database of visible wavelength iris images captured on-the-move and at-a-distance. IEEE Trans. Pattern Anal. Mach. Intell. (2009). doi:10.1016/j.imavis.2009.03.003

Fractal Video Coding Using Modified Three-step Search Algorithm for Block-matching Motion Estimation

Shailesh D. Kamble, Nileshsingh V. Thakur, Latesh G. Malik and Preeti R. Bajaj

Abstract The major problem with fractal-based coding technique is that, it requires more computation at the encoding phase. Therefore, to reduce this computation, block-matching motion estimation algorithms are used. In this paper, we proposed a modified three-step search algorithm (MTSS) for block-matching motion estimation which consists of two cross-search patterns as a few initial step of search, and two cross-hexagon search patterns as a subsequent step of search that is used to search the center part of search window. Experimentations are carried out on standard video databases, i.e., football, flower, akiyo, coast guard, and traffic. The results of efficient three-step search and cross-hexagon search are compared with our proposed approach with respect to the parameters such as mean of absolute difference (MAD) and average search points per frame. Along with this our proposed approach, frame-based technique is used for fractal video coding evaluated on parameters such as compression ratio, PSNR, encoding time, and bit rate.

S.D. Kamble (✉)
Yeshwantrao Chavan College of Engineering, Computer Science and Engineering, Maharashtra, Nagpur, India
e-mail: shailesh_2kin@rediffmail.com

N.V. Thakur
Ram Meghe College of Engineering and Management, Computer Science and Engineering, Badnera-Amravati, Maharashtra, India
e-mail: thakurnisvis@rediffmail.com

L.G. Malik
G.H. Raisoni College of Engineering, Computer Science and Engineering, Nagpur, Maharashtra, India
e-mail: latesh.malik@raisoni.net

P.R. Bajaj
G.H. Raisoni College of Engineering, Nagpur, Maharashtra, India
e-mail: principal.ghrce@raisoni.net

I.K. Sethi (ed.), *Computational Vision and Robotics*, Advances in Intelligent Systems and Computing 332, DOI 10.1007/978-81-322-2196-8_18

Keywords Fractal coding · Motion estimation · MTSS · Three-step search · Cross-hexagon search · MAD · Video coding · Compression ratio · PSNR

1 Introduction

Video compression deals with the compression mechanism for the series of image sequences. In coding, the correlation between the adjacent image frames may get explored, as well as the relativity between them may also be used in the development of the compression mechanism. Generally, the adjacent image frame does not differ much. The probable difference lies in the displacement of the object in the given image frame with respect to the previous image frame. Gray and color videos (i.e., image sequences of gray or color image frame) are the customers for the video compression approach. Different color spaces [1] can be used for the video images from processing point of view. With the very fast development in multimedia communication with moving video pictures, processing on a color images plays a very important role. A color image is represented by 24 bits/pixel in RGB color space format with each color component is represented by 8 bits.

Block-matching motion estimation is a most popular technique for many motion-compensated video coding standards. Video compression standards, in general, are used for the video coding. Basic video compression standards are—Video coding standards are related to the organizations are—ITU-T Rec. H.261, ITU-T Rec. H.263, ISO/IEC MPEG-1, ISO/IEC MPEG-2, ISO/IEC MPEG-4, and recent progress is H.264/AVC. In a series of image sequence, there are spatial, temporal, and statistical data redundancy arises between frames. Motion estimation and compensation are used to reduce temporal redundancy between successive frames. Motion estimation computes the motion/movement of an object in a given image. For achieving data compression in a sequence of images, the motion compensation uses the knowledge of the motion object. Different searching techniques are available to compute the motion estimation between frames. A block-matching motion estimation technology [2, 3] can be used to improve the encoding speed and the compression quality [4].

2 Related Work

Block-matching motion estimation is a vital process for many motion-compensated and video coding standards. Motion estimation could be very computational intensive and can consume up to 60–80 % of computational power of the encode process [5]. So, research on efficient and fast motion estimation algorithms is significant. Block-matching algorithms are used widely because they are simple and easy to be applied. In the last two decades, many block-matching algorithms (BMA) are proposed for alleviating the heavy computations consumed by the brute-force full search algorithm which has the best prediction accuracy, such as the three-step search [6], 2D logarithmic search [7], orthogonal search [8], cross-search [9], binary search [10], new three-step search [11], the four-step search [12], the block-based gradient descent search [13], the diamond search [14], the cross-diamond search [15], and efficient three-step search [16]. All these searches employ rectangular search patterns of different sizes to fit the center-biased motion vector (MV) distribution characteristics [17, 18].

Hexagon-based search employs a hexagon-shaped pattern and results in fewer search points with similar distortion [19]. Block-matching algorithm called novel cross-hexagon search algorithm is proposed in [3]. It uses small cross-shaped search patterns in the first two steps before the hexagon-based search and the proposed halfway stop technique [2]. It results in higher motion estimation speed on searching stationary and quasi-stationary blocks. The traditional algorithms use all the pixels of the block to calculate the distortions that result in heavy computations. Modified Partial Distortion Criterion [20] that uses certain pixels of the block, which alleviates the computations and has similar distortion can be used. New cross-hexagon search algorithm (NHEXS) proposed in [21, 22] consists of two cross-search patterns and hexagon search patterns which is similar in [3] for fast block-matching motion estimation. These search technique is a frame-based fractal video compression technique help to reduce the encoding time and increases the compression quality in fractal coding.

Motion estimation used in the area of video application such that video segmentation, object/video tracking, and video compression. Motion estimation means displacement of pixels position from one frame to another frame which gives the best MV. For estimating a motion in a video sequence, BMA are widely used in most of the video coding standards. In BMA, a frame is divided into a non-overlapping block and for that block MV is estimated. A MV is computed by finding best suitable matched block between previous frame-f and the next frame-$f + 1$ as shown in Fig. 1. MV for the block $B_f(x, y)$ is computed as $(+1, -1)$ i.e., MV-$B_f(x,y) = (+1, -1)$.

Still research is going on for efficient and fast block-matching motion estimation. The rest of the paper is organized as follows. Section 3 reviews three-step search algorithm. Section 4 describes efficient three-step search algorithm. Section 5 describes cross-hexagonal search algorithm. Proposed approach is presented in Sect. 6. Section 7 consists of simulation results. Finally, Sect. 8 presents the conclusion based on proposed work.

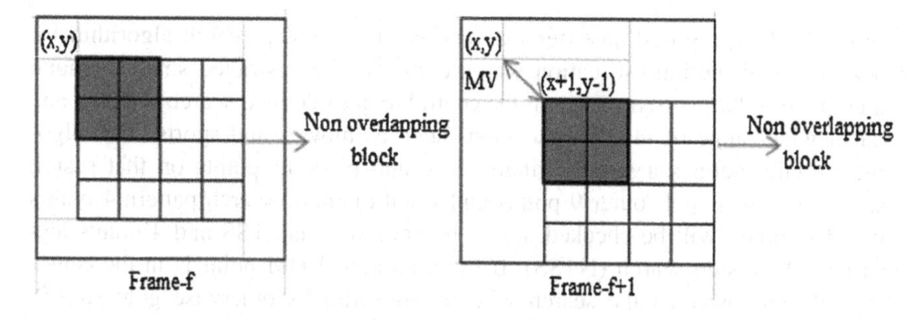

Fig. 1 Block-matching motion vector estimation

Fig. 2 Three-step search motion estimation

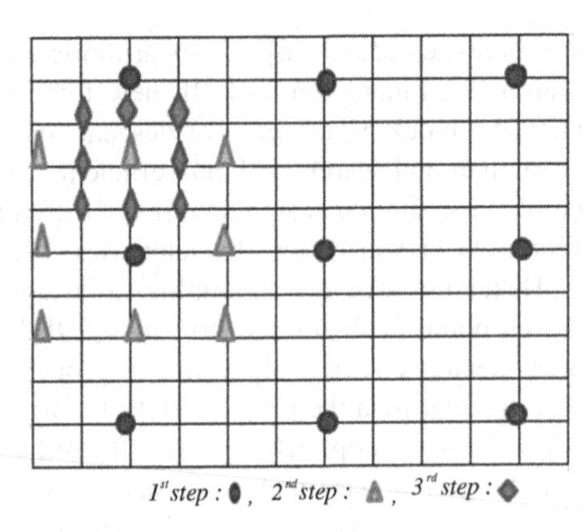

3 Three-step Search Algorithm

Koga et al. [6] proposed a three-step search motion estimation algorithm which is simple in nature and easy to implement. This algorithm is one of the popular BMA for the low bit rate application because of its performance based on the number of checking points in the search window. Computational cost of this algorithm is low as compared to the full search algorithm. First step is to define a search window size for searching the best match. In the first step, plot nine points in the search window at the equal distance of step size. In the second step, the step sizes are divided by 2 if minimum block distortion measure (BDM) point is one of the nine points of search window and consider this point as a center point in the third step. Same process in step two is repeated again in step three. For larger search window, TSS can be easily extended up to n-steps until the step size is smaller than one. This complete process is shown in Fig. 2. If the step size is $s = 7$ then the number of checking points required for TSS is 25 using equation, i.e., number of checking points $= 1 + 8 \log_2 (s + 1)$ where s is the step size of search window.

4 Efficient Three-step Search Algorithm

Jing et al. [16] proposed an extended version of three-step search algorithm for block-matching motion estimation which consists of unrestricted small diamond search pattern that is used to search the central area of defined search window and used in wide range of video applications such as movies and sports. This algorithm initially defines a search window size and plots 13 points on that search window size. In step 1, outer 9 points and small diamond search pattern 4 points (Total 13 points) will be checked, i.e., 4 points more than TSS and 4 points less than new three-step search (NTSS). If the minimum BDM point is in the center of search window then the search will be stopped/halts otherwise goto step 2.

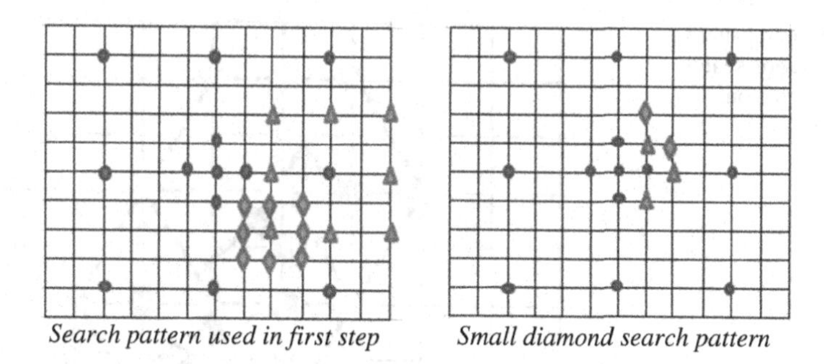

Fig. 3 Efficient three-step search motion estimation

In step 2, if minimum BDM point is one of the outer 8 points then 3SS algorithms are used to search the point otherwise goto step 3. In step 3, if minimum BDM point is one of the four points on the small diamond pattern then consider this point as a center and checked another 3 points. This process is repeated until small diamond search pattern reaches to search window boundary. In this algorithm, the number of check points required for ETSS is 13 for best case, 13–29 for average case, and $9 + 4 + 8 + 8 = 29$ for worst case as shown in Fig. 3.

5 Cross-hexagonal Search Algorithm

Belloulata et al. [22] introduces a new cross-hexagonal block-matching motion estimation search algorithm consisting of two cross-search patterns i.e., a small and large cross-shaped patterns as a few initial steps of search and two cross-hexagon search patterns, i.e., small and large hexagon search patterns as a subsequent steps of search as shown in Fig. 4. This algorithm initially starts with small cross-shaped pattern consisting of 5 points located at the center of search window. If minimum BDM point found at the center of the small cross-shaped pattern then stop searching, i.e., the number of checking point required for new cross-hexagonal search is 5 for best case which is better than all the techniques available for estimating a MV otherwise consider a minimum BDM point as a center of newly formed small cross-shaped pattern. If minimum BDM point found at the center of newly formed small shape pattern then stop searching, i.e., another best case solution is 8 checking points are required for finding out best possible MV, otherwise check the another 3 unchecked points of large cross-pattern and 2 unchecked points of the square center biased to show the best possible direction for the hexagonal search. A new large hexagonal-shaped pattern is formed by considering a center point as minimum BDM point found in small cross-search pattern. If minimum point found at the center of large hexagonal pattern then large hexagonal pattern shifted/changed to small hexagonal pattern and find best MV in

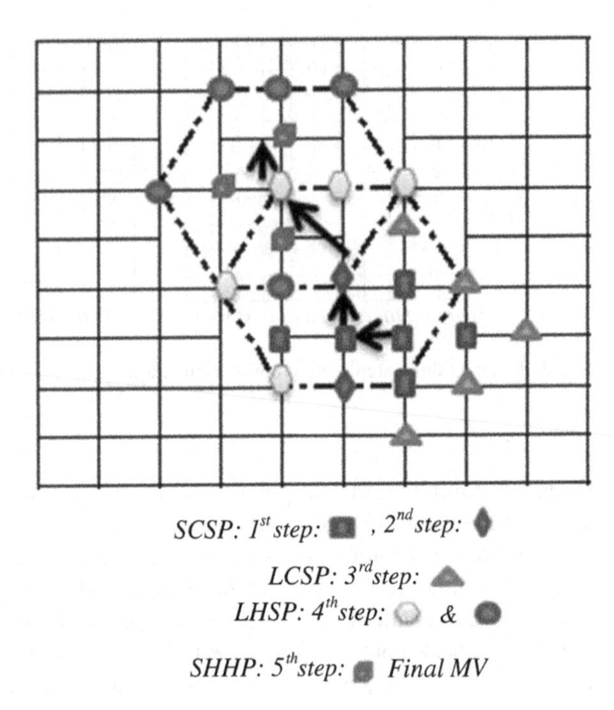

Fig. 4 Search pattern in new hexagon search motion estimation

small hexagon-shaped pattern, otherwise again form a new large hexagon pattern; this formation of new large hexagon pattern is repeated until minimum BDM point found at the center of large hexagonal pattern.

6 Proposed Approach/Algorithm

Two cross-search patterns, i.e., small and large cross-search patterns and two cross-hexagon search patterns, i.e., large cross-hexagon and small cross-hexagon search patterns are used in the center part of search window to exploit central-biased characteristic of MV in video sequences. Figure 5 shows the search pattern used in first step of proposed approach.

In first step, total 9 + 4 points are checked out of 17 checking points. If minimum BDM point found at the center of 9 + 4 points then stop searching otherwise go to second step. If minimum BDM point found at the outer part of search window then search process is same as TSS otherwise go to third step. If minimum BDM point found at the 4 outer points of small cross-search pattern then search process is same as cross-hexagon search. There is no restriction on searching in the center window part unless minimum BDM point found at the center of large cross-hexagon pattern or large cross-hexagon search pattern reaches to the

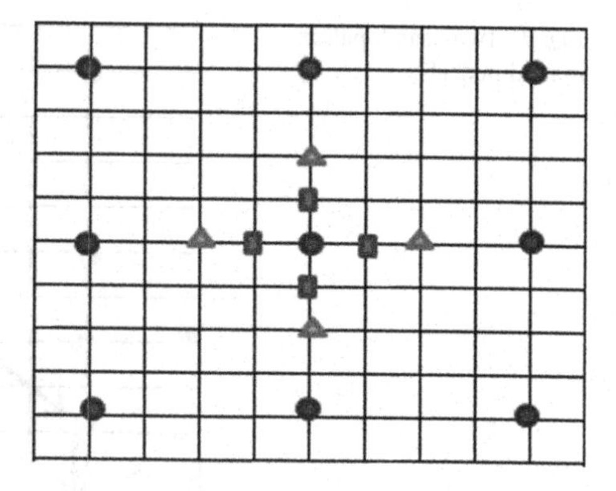

Fig. 5 Search pattern used in first step of modified TSS

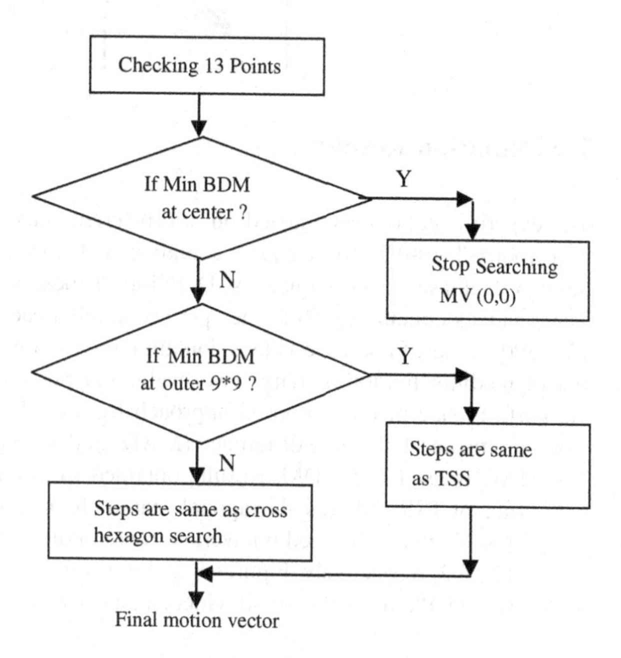

Fig. 6 Block schematic of the modified TSS

outer boundary of the search window. This unrestricted search in the central part of window increases the probability of finding a true MV within the center area of window. Block schematic and search pattern of proposed modified approach are shown in Figs. 6 and 7, respectively.

The number of checking points required for this algorithm is 13 for best case i.e., stationary block, 13–29 points are required in the central 5×5 area of search window i.e., average case, and $9 + 4 + 8 + 8 = 29$ points are required for the worst case. These checking points are compared with the NTSS algorithm and are 17 for best case, 17–33 points for average case, and 33 checking points for worst case.

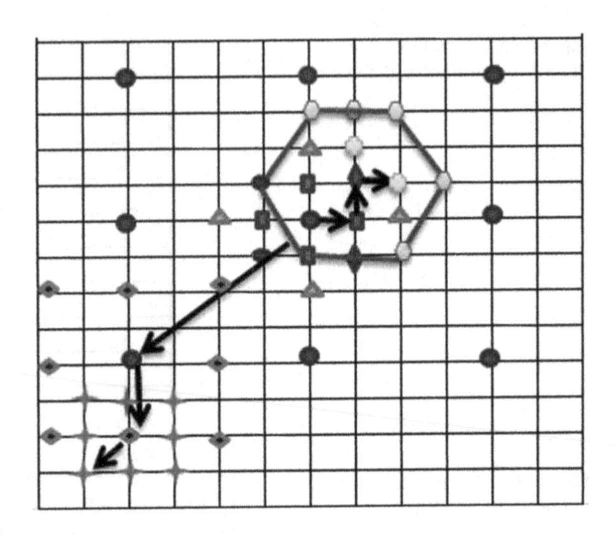

Fig. 7 Two-search pattern in modified TSS

7 Simulation Results

The experimentations are carried out on different standard video sequences database football, traffic, coast guard, container and flower video, and also on user-defined database nature sequences. Duration of these videos is from 8 s. to 10 s. These videos contain 15–30 frames per second and each video sequence contains 120–300 frames. First three videos contain large motion activity while others consist of medium motion activity. Two evaluation parameters are used to measure the performance of our proposed approach by considering 50 frames from each video are mean of absolute difference (MAD) and average search point per frame. The MAD is used as a BDM. Results obtained from our approach are compared with efficient TSS and cross-hexagonal search algorithm. Performance analysis of our proposed approach based on average search point and MAD is summarized in Tables 1 and 2, respectively. Figures 8 and 9 illustrate the frame by frame in terms of MAD and PSNR for football video, respectively. Figure 10 shows decoded

Table 1 Performance analysis of MTSS, E3SS, and CHS

Video name	No. of average search points per frame		
	MTSS	E3SS	CHS
Football	601.2345	5.9923e + 03	6.741e + 03
Flower	601.1935	5.6359e + 03	7.3924e + 03
Akiyo	543.6734	6.1481e + 03	7.2453e + 03
Coast guard	563.3591	5.4614e + 03	4.5039e + 03
Container	93.3814	4.9487e + 03	4.9667e + 03
Mobile	651.3521	5.3084e + 03	6.3061e + 03
Template	678.1932	4.9255e + 03	4.6833e + 03
Nature	601.1357	5.5557e + 03	7.4212e + 03
Traffic	601.1935	5.0127e + 03	5.3421e + 03

Table 2 Performances analysis of MTSS, E3SS, and CHS by using MAD

Name of video	No. of computation required		
	MTSS	E3SS	CHS
Football	7.0348	25.5373	29.1130
Flower	7.2983	21.1139	27.6289
Akiyo	6.3478	19.4534	24.8965
Coast guard	7.0318	26.4212	25.1250
Container	7.3158	23.9874	28.4688
Mobile	6.7082	23.4603	29.8359
Template	7.0318	24.7103	29.8672
Nature	6.4989	23.6754	28.9576
Traffic	6.6889	26.2909	31.6563

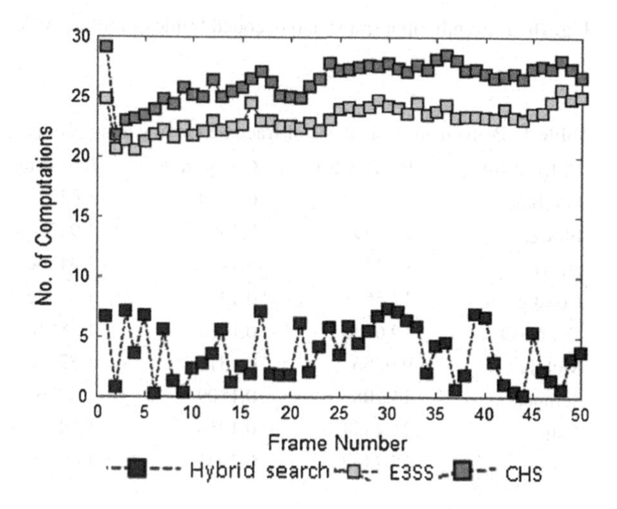

Fig. 8 Computational complexity of football video by using MAD

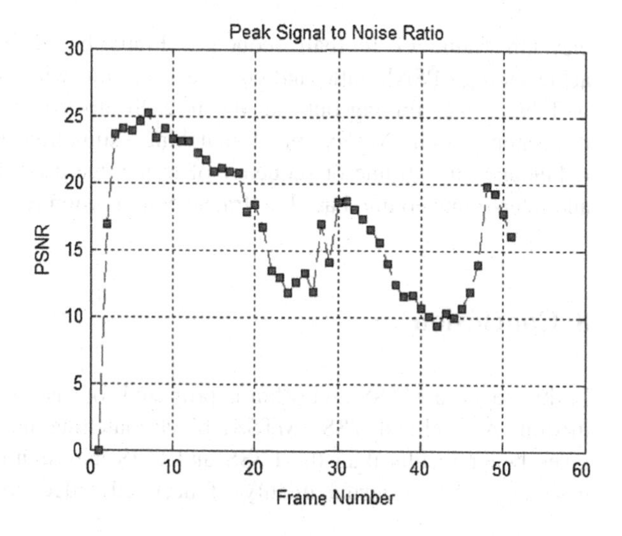

Fig. 9 PSNR graph for football video

Fig. 10 Eleventh input frame and decoded frame of football video

Table 3 Performances analysis of fractal encoding and decoding

Video name	PSNR (db)	Comp. ratio	Encoding time (s)	Bit rate (Kbps)
Football	25.0162	0.1614	1.6352e + 003	3.9408e + 003
Flower	11.2397	0.1528	1.9452e + 003	3.9408e + 003
Akiyo	17.3478	0.1489	1.3456e + 003	3.9408e + 003
Coast guard	17.8661	0.1381	1.6333e + 003	3.9408e + 003
Container	19.6230	0.1487	1.7851e + 003	3.9408e + 003
Mobile	9.0256	0.1310	1.7223e + 003	3.9408e + 003
Tempete	14.7083	0.1509	1.5826e + 003	3.9408e + 003
Nature	21.4374	0.1259	1.5468e + 003	3.9408e + 003
Traffic	25.2432	0.2941	1.4189e + 003	3.9408e + 003

eleventh frame of football sequence. Frame-based fractal video compression achieves high PSNR ratio and compression ratio with good image quality shown in Table 3. For finding out the MV in each block in frame, we used a modified three-step search (MTSS) block-matching estimation algorithm. To remove the redundancy in a frame or sequence of frame, we used the concept of intra-frame and inter-frame coding based on fractal compression technique.

8 Conclusion

In this paper, a MTSS algorithm is proposed for fast block-matching motion estimation. A modified TSS (MTSS) block-matching motion estimation approach gives better results than the E3SS and CHS and significantly larger PSNR values, i.e., a better visual quality of decoded video compared with the results.

The performance comparison on the number of search points per block of MTSS is similar to the ETSS. MTSS performs efficiently for small and large motion estimation because of combination of two approaches, i.e., TSS and CHS. Hence, the MTSS algorithm is used in video applications. In future, MTSS approach combines with the process of weighted finite automata-based inter-frame coding for obtaining MVs in a block.

References

1. Gonzalez, R.C., Woods, R.E.: Digital Image Processing, vol. 2. Pearson Education Asia, Hong Kong (2005)
2. Zhu, S., Tian, J., Shen, X., Belloulata, K.: A new cross-diamond search algorithm for fast block motion estimation. In: IEEE International Conference on Image Processing, ICIP'09, Cairo, vol. 1, pp. 1581–1584, 07–11 Nov 2009
3. Zhu, S., Tian, J., Shen, X., Belloulata, K.: A novel cross- hexagon search algorithm based on motion vector field prediction. In: IEEE International Symposium on Industrial Electronics, ISIE'09, Seoul, Korea, pp. 1870–1874, July 2009
4. Zhu, S., Hou, Y., Wang, Z., Belloulata, K.: A novel fractal video coding algorithm using fast block matching motion estimation technology. In: International Conference on Computer Application and System Modeling, ICCASM'10, Taiyuan, China, vol. 8, pp. 360–364, Oct 2010
5. Information Technology-Coding of Audio Visual Objects-Part 2: Visual, ISO/IEC 14 469-2 (MPEG-4 Visual), 1999
6. Koga, T., Iinuma, K., Hirano, A., Iijima, Y., Ishiguro, T.: Motion compensated interframe coding for video conferencing. In: Proceedings of National Telecommunications Conference, New Orleans, LA, pp. G5.3.1–G5.3.5 (1981)
7. Jain, J.R., Jain, A.K.: Displacement measurement and its application in interframe image coding. IEEE Trans. Commun. **29**, 1799–1808 (1981)
8. Puri, A., Hang, H.M., Schilling, D.L.: An efficient block matching algorithm for motion compensated coding. In: Proceedings of IEEE International Conference on Acoustics, Speech, and Signal Processing, pp. 1063–1066, (1987)
9. Ghanbar, M.: The cross search algorithm for motion estimation. IEEE Trans Commun **38**, 950–953 (1990)
10. Urabe, T., Afzal, H., Ho, G., Pancha, P., El Zarki, M.: MPEG Tool: an X window-based MPEG encoder and statistics tool. Multimedia syst **1**(5), 220–229 (1994)
11. Li, R., Zeng, B., Liou, M.L.: A new three-step search algorithm for block motion estimation. IEEE Trans. Circuits Syst. Video Technol. **4**(4), 438–443 (1994)
12. Po, L.-M., Ma, W.-C.: A novel four-step search algorithm for fast block motion estimation. IEEE Trans. Circuits Syst. Video Technol. **6**(3), 313–317 (1996)
13. Liu, L.K., Feig, E.: A block based gradient descent search algorithm for block motion estimation in video coding. IEEE Trans. Circuits Syst. Video Technol. **6**(4), 419–422 (1996)
14. Zhu, S., Ma, K.K.: A new diamond search algorithm for fast block-matching motion estimation. IEEE Trans. Image Process. **9**(2), 287–290 (2000)
15. Cheung, C.H., Po, L.M.: A novel cross-diamond search algorithm for fast block motion estimation. IEEE Trans. Circuits Syst. Video Technol. **12**(12), 1168–1177 (2002)
16. Jing, Xuan, Lap-Pui, Chau: An efficient three-step search algorithm for block motion estimation. IEEE Trans. Multimedia **6**(3), 435–438 (2004)
17. Tham, J.Y., Ranganath, S., Ranganath, M., Kassim, A.A.: A novel unrestricted center-biased diamond search algorithm for block motion estimation. IEEE Trans. Circuits Syst. Video Technol. **8**(4), 369–377 (1998)

18. Chen, H.M., Chen, P.H., Yeh, K.L., Fang, W.H., Shie, M.C., Lai, F.: Center of mass-based adaptive fast block motion estimation. EURASIP J. Image Video Process. vol. 2007, Article ID 65242, 11 p. (2007)
19. Zhu, C., Lin, X., Chau, L.-P.: Hexagon-based search pattern for fast block motion estimation. IEEE Trans. Circ. Syst. Video Technol. **12**, 349–355 (2002)
20. Cheung, C.H., Po, L.M.: Normalized partial distortion search algorithm for block motion estimation. IEEE Trans. Circ. Syst. Video Technol. **10**, 417–422 (2000)
21. Hong-ye, L., Ming-jun, L.: Cross-Hexagon-based motion estimation algorithm using motion vector adaptive search technique. In: International Conference on Wireless Communications and Signal Processing, (2009)
22. Belloulata, K., Zhu, S., Wang, Z.: A fast fractal video coding algorithm using cross-hexagon search for block motion estimation. International Scholarly Research Network Signal Processing, vol. 2011, Article ID 386128, (2011)

Designing and Development of an Autonomous Solar-Powered Robotic Vehicle

Pankaj P. Tekale and S.R. Nagaraja

Abstract In this paper, design and fabrication of a solar-powered robotic vehicle is presented. The energy for the vehicle is supplied by two solar panels of 5 W each. For efficient energy management, a charging system is designed. The charging is independent of the vehicle movement. Two batteries are used so that one is charging, while other is discharging. A charge controller is also designed so as to provide direct power supply to the connected load in case both batteries fail.

Keywords Li-Io batteries · Solar panels · Robotic vehicle

1 Introduction

The Sojourner is the first rover landed on the Mars on 4 July 1997, which is the first solar-powered robotic vehicle [1]. Since the 1960s, there have been efforts worldwide to develop robotic mobile vehicles for traversing planetary surfaces. Developments in mobility, navigation, power, computation, and thermal control. Many problems faced by the past rover are solved by the present rovers [2]. Rechargeable batteries in a space mission were used for the first time in the Mars exploration rover [3]. Field-integrated design and operations (FIDO) rover, an advanced NASA technology development platform and research prototype, was used for the NASA's rover mission to Mars [4]. NASA has developed the different kind of generation of Mars exploration rover such as

P.P. Tekale (✉)
Embedded System, Amrita Vishwa Vidyapeetham, Bangalore Campus, Bangalore, India
e-mail: pptekale28@gmail.com

S.R. Nagaraja
Department of Mechanical Engineering, Amrita Vishwa Vidyapeetham, Bangalore Campus, Bangalore, India
e-mail: sr_nagaraja@blr.amrita.edu

I.K. Sethi (ed.), *Computational Vision and Robotics*, Advances in Intelligent Systems and Computing 332, DOI 10.1007/978-81-322-2196-8_19

k9, MER, Spirit, and curiosity [4, 5]. Solar radiation is radiant energy emitted by the sun, particularly electromagnetic energy. It is also known as short-wave radiation. Solar radiation comes in many forms, such as visible light, radio waves, heat (infrared), X-rays, and ultraviolet rays. The sun is the Earth's major energy source and radiates its energy from a distance of 150 million kilometers, or 8.3 light minutes. This solar radiation reaches the outside of our atmosphere with an irradiance of about 1,360 W/m^2. It covers the spectrum from ultraviolet, through visible, to near infrared wavelengths. Solar radiation is very important factor for space missions and it is measured in Sieverts (sV). Whereas Martian atmosphere is very thin at around 1% the density of Earth's air and no magnetosphere. The total solar irradiation on Mars is 475 W/m^2 [6].The proposed autonomous solar-powered robotic vehicle consists of the robotic platform, battery switching system, battery charging and discharging system, and battery monitoring system.

2 Mobile Robotic Platform

Solar-powered robotic vehicle is four-wheeled independent suspension mechanism vehicle. It consists of plastic plate (used as chassis) of 4 mm thickness, two polycrystalline solar panels of 5 W each and fixed to the plastic plate using threaded rod (used as spacer). Each wheel consists of one dc motor for forward and backward movement of vehicle with 100 rotation per minute, and its torque is 30 kg/cm. Robotic vehicle is compact, reduced weight, and small dimension makes the vehicle ideal as exploration vehicle (Fig. 1).

The programming part of the vehicle is mainly carried out in a C language. The other level language is assembly language as vehicle consists of microcontroller (Arduino development board) to control all function of the vehicle.

Fig. 1 Solar-powered robotic vehicle

3 Mechatronics System

The power management system consists of a pack of two batteries charged using solar panels, and the whole things are controlled by microcontroller that why it make the smart power management system. But the concept of the smart is applied to the software part. Our aim is to build the low cost and efficient power management system for simple pack of two batteries. The system consists of solar panels, charge controller, path selector etc.

The microcontroller controls the consumption and decision making of the vehicle. The aim of the power management system is to track maximum solar energy and taking the data from batteries and solar panels, and controlling the charging (Fig. 2).

3.1 Batteries Switching System

The switching system consists of single channel 5VDC Double pole double throw (DPDT) relay board with break-before-make operation logic. Their function is connecting electrically the charge and discharge paths between the batteries. The batteries are connected to the normally open (NO) and normally closed (NC) terminal of the relay. The relay board is controlled by microcontroller board.

3.2 Charging and Discharging System

Charging and discharging system of the autonomous vehicle consists of a PWM charge controller, which will manage the charging and discharging path between the batteries. Figure 3 shows the block diagram of the power management of the vehicle. Solar panels, batteries, and load (microcontroller board) is connected to the charge controller at their respective terminals.

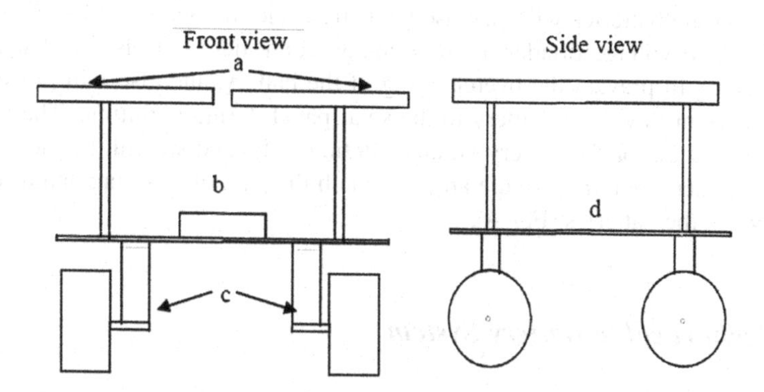

Fig. 2 Mechanical design of solar-powered robotic vehicle: **a** solar panels, **b** batteries and controlling hardware, **c** suspension mechanism and dc motor, and **d** plastic plate as chassis

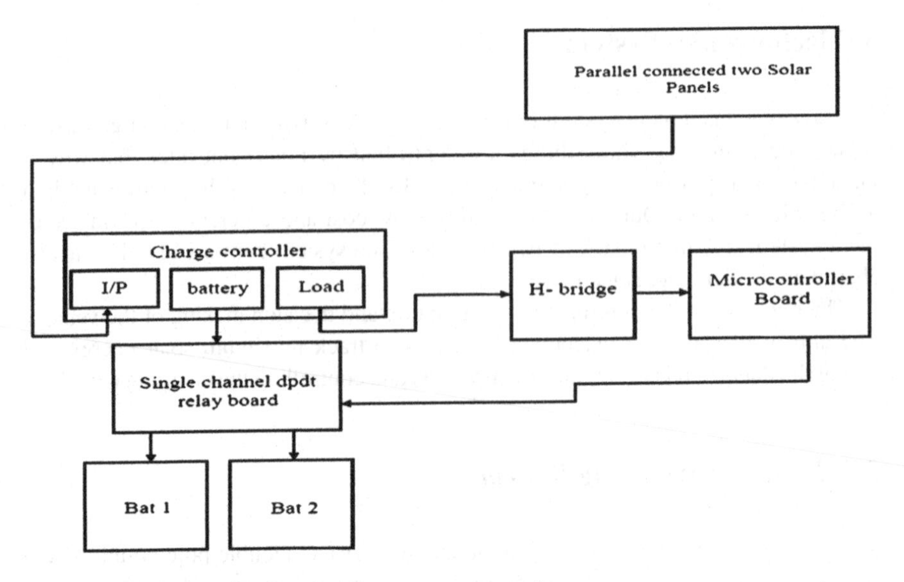

Fig. 3 Overall connection diagram of energy management

During sunny days (good solar radiation), charge controller will charge the batteries also provide the power to the load simultaneously. While in cloudy days (bad solar radiation), charge controller will provide power to the load through the battery. So, because of these characteristics of the charge controller, we can improve and increase the life cycle of the batteries.

Charge controller is also used to prevent the batteries from over charging and reverse current.

3.3 Batteries Monitoring System

Monitoring the batteries will increase their life cycle. It consists of battery charge controller and voltage divider. Battery charge controller controls the charging of the battery, will prevent the overcharging of the battery, and also will prevent the reverse current flow from battery to the solar panel during nighttime. The voltage divider will monitor the battery voltage continuously, and so with the help of the voltage divider's reading, we are able to switch the path of charging and discharging between the batteries (Fig. 4).

3.4 Rechargeable Battery System

The PV system is used to charge the pair of battery in such fashion that while one battery charges the other battery will discharge at the same time. The conventional

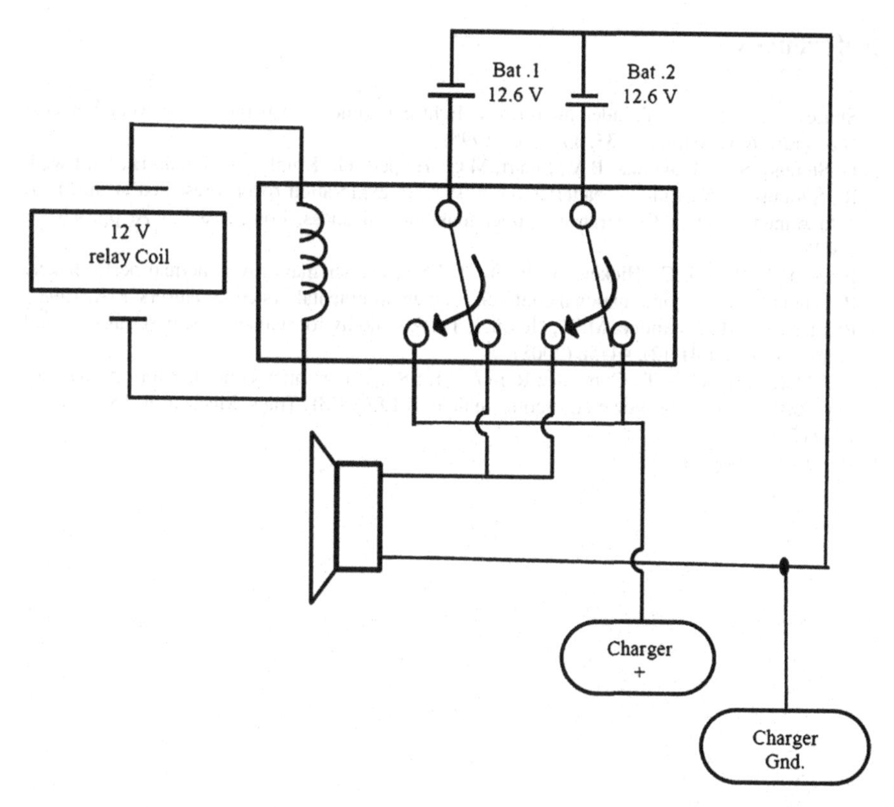

Fig. 4 Overall connection of switching system of the batteries

robotic vehicle uses single battery to provide energy; hence, the robotic vehicle remains in idle state until battery is charged.

The battery charge controller makes the direct connection between the load and PV system when both the batteries are unable to provide the current to the vehicle.

4 Conclusion

This paper has presented an energy management system applied to an autonomous robotic vehicle. The proposed switching mechanism charges two batteries alternatively, hence amount of generated power is independent from the vehicles mobility. So, solar panels can power a single battery at a time. This design does not attempt to address high charging times or longer operating times but to prove sustainable and feasible solution for a robotic vehicle. The main advantage is that if both batteries fail, charge controller provides direct power to the load.

References

1. Shirley, D.L.: Mars pathfinder microrover flight experiment—a paradigm for very low-cost spacecraft. Acta Astronaut. **35**, 355–365 (1995)
2. Di Stefano, S., Ratnakumar, B.V., Smart, M.C., Halpert, G., Kindler, A., Frank, H., Di, Ewell, R., Surampudi, S.: Lithium batteries on 2003 mars exploration rover. Presented at the IEEE 17th Annual. Battery Conference Applications and Advances, Long Beach, CA, pp. 47–51, 2002
3. Eisen, H.J., Wen, L.C., Hickey, G., Braun, D.F.: Sojourner mars rover thermal performance. Presented at the 28th international conference on environmental systems, Danvers, MA, 1998
4. Bajracharya, M., Maimone, M.W., Helmick, D.: Autonomy for mars rovers: past, present, and future. Computer **41**(12), 44–50 (2008)
5. de J Mateo Sanguino, T., Gonz´alez Ramos, J.E.: Smart host microcontroller for optimal battery charging in a solar-powered robotic vehicle. IEEE/ASME Trans. Mechatron. **18**(3):1039–1079 (2013)
6. http://www.google.co.in/

SONIC: Stability of Nodes in Cooperative Caching for Wireless Network

Vivek Kumar, Narayan Chaturvedi and Prashant Kumar

Abstract Most of the researches in wireless networks focus on data transfer through central node, and not much works have been done on data access, cache replacement, and stability of nodes. A general technique used to improve the performance of data access is cooperative caching. Cooperative caching has become one of the most exciting techniques for wireless environment. Cooperative caching allows sharing and coordinated cached data items which are stored in node's local cache to achieve better performance. The main strength of this technique is lower node communication overhead and energy consumption as well as reduction in the query latency. Cooperative caching caches the frequently accessed data, the nodes do not always have to send request to server. But, high node mobility, constrained battery power, and limited wireless bandwidth may decrease the cache hit ratio and increase access latency or nodes in a region may fail to cache more hot data when a cooperative caching scheme cannot effectively distribute varied data to cache in that region. Nodes in wireless network may have similar task and share general interest, cooperative caching which allows the sharing and coordination of cached data among multiple nodes can be used to reduce the bandwidth and power consumption. In this paper, we propose the SONIC scheme; we find the *stability of nodes*, and the nodes which are more stable are being used first for searching the data, thereby increasing the network lifetime.

V. Kumar (✉) · N. Chaturvedi
Department of Computer Science and Engineering, Graphic Era University, Dehradun, India
e-mail: vivekror7@gmail.com

N. Chaturvedi
e-mail: narayanchaturvedi@gmail.com

P. Kumar
Department of Computer Science and Engineering, National Institute of Technology, Hamirpur, India
e-mail: prashantkumar32@gmail.com

I.K. Sethi (ed.), *Computational Vision and Robotics*, Advances in Intelligent Systems and Computing 332, DOI 10.1007/978-81-322-2196-8_20

Keywords Cooperative cache · Wireless network · Cache replacement · Cache placement

1 Introduction

Cooperative caching has become one of the most exciting techniques in mobile environment. Cooperative caching allows sharing and coordinating cached data items, which are stored in mobile client's local cache to achieve a better performance. This technique has got lower mobile host communication overhead and energy consumption as well as reduces the query latency. Cooperatively caching frequently accessed information, mobile clients do not always have to send requests to server. But, high node mobility, restricted battery energy, and limited wireless bandwidth may decrease the cache hit ratios and increase access latency or nodes in a region may fail to cache more hot data when a cooperative caching scheme cannot effectively distribute varied data to caches in that region. As mobile nodes in ad hoc networks may have similar tasks and share general interest, cooperative caching that allows the sharing and coordination of cached data among multiple nodes can be used to reduce the bandwidth and power utilization [1].

Wireless communication is the transfer of information between two or more points that is not connected by electrical conductor. Wireless networking is used to meet needs. The most common use is to connect nodes that travel from location to location. Another common use is for mobile network that connect via base station, satellite. Wireless mobile communication is a fastest growing segment in communication industry [2].

A lot of researches have been done to improve the caching performance in wireless environment. The two basic types of cache sharing techniques are push based and pull based. Push-based caching technique: When a node acquires and caches new data item, it actively advertises the caching event to the node in its neighborhood. Mobile node in the vicinity will record the caching information upon receiving such an advertisement and use it to direct subsequent request for the same item. This scheme enhances the usefulness of the cached contents. But this introduction communication overhead for the advertisement and further an advertisement may become useless if no demand for the cached item arise in the neighborhood.

In this push-based scheme, the caching information known to a node may become obsolete due to node mobility or cache replacement. This problem is resolved in pull-based approach. Pull-based caching technique: When a mobile node wants to access a data item that is not cached locally, it will broadcast a request to the nodes in its vicinity. A nearby node that has cached the data will send a copy of the data to the request originator. So, pull-based caching techniques allow the node to utilize the latest cache contents. However, as compare to the pushing technique, the pulling scheme has two demerits:

In case the request data item is not cached by any node in the neighborhood, the request node will wait for the time-out interval to expire before it proceeds to send another request to the data center. This will cause extra access latency, and the pulling effort is in vain.

Pulling technique uses broadcast to locate a cached copy of an item so it may happens that more than one copy will be returned to the requester node if multiple nodes in the neighborhood cache the requested data. This introduces extra communication overhead [3].

2 Motivation

There are several issues associated with existing cooperative caching schemes or algorithm that triggered a motivation to do the current research. The design of caching strategy in wireless network benefits from the assumption of existing end to end paths among mobile nodes and the path from a requester to the data sources remains unchanged during the data access in most cases. Such assumption enables any intermediate node on the path to cache the pass-by data. One of the interesting problems is an efficient use of a cache memory and data sharing with neighbor node. The idea is to store as many blocks as possible in the cache without increasing its size because the cache that contains more blocks would have hit rate. The existing schemes or algorithms reach this objective at the cost of large global cache memory. Yet, the more clients are connected to the cooperative caching system, the more memory is consumed space by the global cache. Biggest problem in cooperative caching we are not much data store in cache some problem also creates cache replacement and placement data from cache memory. SONIC scheme is to resolve this issue; we find the *stability of nodes*, and the nodes which are more stable are being used first for searching the data, thereby increasing the network lifetime.

3 Organization of Paper

The rest of this paper is organized as follows: Section 4 describes the literature review related to the cooperative cache. Section 5 describes the proposed work. Section 6 presents the simulation and results. Finally, Section 7 concludes the research work and provides a brief insight to the possible future work.

4 Related Work

Gao and Srivatsa [4] proposes a novel scheme to address the aforesaid challenges and to efficiently support cooperative caching in disruption tolerant networks (DTNs). Their basic idea is to purposely cache data at a set of network central locations (NCLs) each of which corresponds to a group of mobile nodes being easily accessed by other nodes in the network. Each NCL is represented by central node which has high popularity in the network and prioritizes for caching data. Due to the limited caching buffer of central nodes multiple nodes near a central node may be involved for caching and they develop a capable approach to NCL selection in DTNs based on a probabilistic selection metric.

Yin and Cao [5] studies the design and evaluates the cooperative caching techniques to efficiently support data access in ad hoc networks. In this paper, first two schemes proposed: Cache Data which caches the data and Cache Path which caches the data path. Firstly analyze both scheme cache data and cache path then after proposed new scheme. The new scheme is hybrid approach. A hybrid approach which can further improve the performance by taking advantage of Cache Data and Cache Path whereas avoiding their weaknesses. Hybrid cache is a combination of cache data and cache path.

Minh and Bich [6] explains that data availability in mobile information systems is lower than in traditional information systems due to its limitations of wireless communication such as easy disconnection, instability, and low bandwidth. Cooperative data caching is an attractive approach to this problem. This paper proposes an efficient approach for cooperative caching in mobile information system. They evaluated and demonstrated the effectiveness of our approach by experiments on simulation datasets. Cooperating caching is more interesting problem in mobile environment.

Zhao and Zhang [7] presents design and implementation of cooperative cache in wireless P2P networks and proposes solutions to find the best place to cache the data. This paper proposes a novel asymmetric cooperative cache approach where the data requests are transmitted to the cache layer on every node but the data replies are only transmitted to the cache layer at the intermediary nodes that need to cache the data. This solution not only reduces the overhead of copying data between the user space and the kernel space but also allows data pipelines to reduce presentation of cooperative cache.

Du et al. [8] proposes a cooperative caching scheme called COOP for MANETs to improve data availability and access performance. COOP addresses two basic problems of cooperative caching. For cache resolution, COOP uses the cocktail approach that consists of two basic schemes: hop-by hop resolution and zone-based resolution, and a least utility value (LUV)-based replacement policy has been used to improve the efficiency of ZC caching [9].

5 Proposed Work

In SONIC scheme, firstly if node requires data, it sends request to all adjacent neighboring nodes. Then these nodes send request to their next neighboring nodes and so on. If all the neighboring nodes fail to send back the required data, then the request is forwarded to the server.

Now the desired data is replaced to that node whose stability is maximum. We have also used some exiting techniques, for example, TTL, data utilization, count access, and label on data, i.e., primary data and secondary data. Data is transferring through TCP protocol and encrypted using 128 bit key which is generated randomly.

S_n Stability Node
P Power
B Bandwidth
S Size
M Mobility.

If a node is more stable, then the power consumption is less on that node.

$$S_n \propto P$$

If any node is at rest, then it has more bandwidth. Its bandwidth utilization is low in comparison with the node which works.

$$S_n \propto B$$

In SONIC scheme, the node which is more stable results in more cache utilization.

$$S_n \propto S$$

and also some other, if node stay one place then other node more data share as compare to the mobility because its maintain the all condition TTL, count access, PD, SD, and table of cache. Stable of node is just opposite of mobility node.

$$S_n \propto \frac{1}{M}$$

Node with more stability then increases the data access and data availability and reduces the power consumption and frequent data access near node.

$$S_n \propto \frac{P * B * S}{M}$$

$$S_n = SN_c \frac{P * B * S}{M}$$

SN_c is Constant

S server
D_e Encrypted Data
D_d Simple Data (Decrypted data)
N_i Sources Nodes

Assumptions

1. 20 nodes for *object 1* and 10 for *object 2* are placed randomly.
2. Each node has equal energy and equal bandwidth.
3. Data pool is created to store transfer data.

6 Algorithm

Step1 for i ← 1 to max
Step2 if sensor id and protocol match then proceed
Step3 D_e—D*128 bit key
Step4 for k ← 1 to max
Step5 S ← Data
Step6 Memory ← update
Step7 If status ← available (require data)
Step8 Send report to N_i
Step9 Update data
Step10 D_d (Reverse process of encryption)
Step11 Reset memory of data pool
Step12 Else
Step13 Put report on hold
Step14 End if
Step15 Else
Step16 Flash warning "unauthorized server found…"
Step17 End if
Step18 End for

Let us consider a wireless sensor networks scenario. This network has fixed infrastructure and nodes which are free to move anywhere in the network. Since nodes are mobile, the topology is dynamic and temporary. There exists a data server that contains the database (Fig. 1).

We have assumed *object 1* for one zone and *object 2* for another zone. Nodes are deployed in each zone randomly. Each node has equal energy and equal bandwidth. Data transfer through TCP protocol. All nodes have equal energy of 0.2. Firstly, we have deployed the nodes randomly in zone then nodes send request to all neighbor nodes if the nodes unable to find that, then it will send its request to data pool. Data transfer is encrypted using 128 bit key and it is generated randomly. Data pool has all the information related to the objects in their respective zones. After getting response, all the nodes in cache memory and data pool update the information. We have implemented our algorithm on a hospital scenario. We have deployed 20 nodes representing the patients and 10 nodes representing the doctors, randomly (Table 1).

The nodes begin to search data; firstly the node requests data from all neighbor nodes. If the node is unable to find data, then it will send its request to data pool. After getting response, all nodes update the cache memory and the data pool memory and the desired data are replaced to that node whose stability is maximum (known as memory balancing) (Fig. 2).

A graph comparing the time required for the caching process between our proposed method (SONIC scheme) and the normal method. It is evident that our method takes less time than the normal method as number of nodes increases (Figs. 3 and 4).

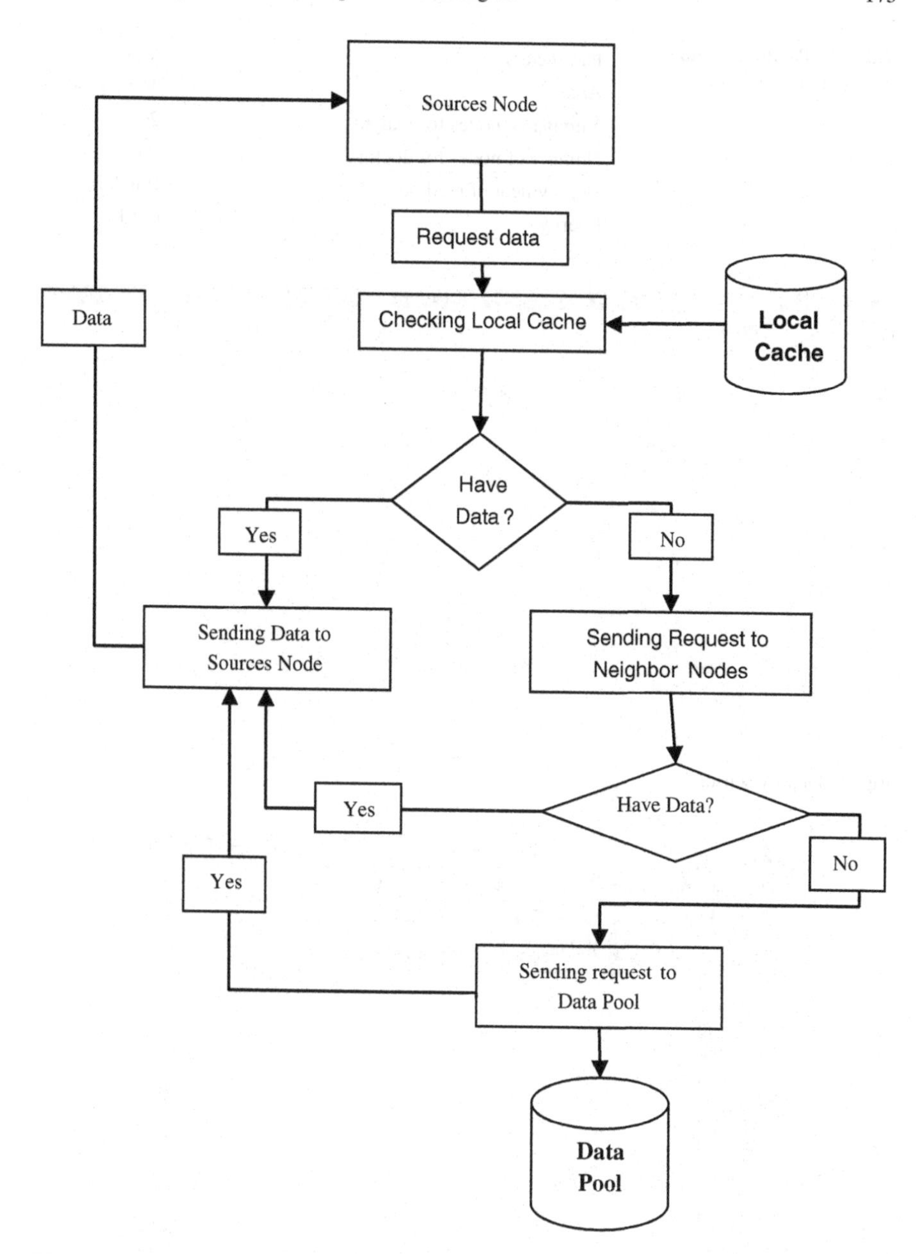

Fig. 1 Flowchart of the proposed algorithm

Table 1 Parameter values used in the simulation

Parameters	Values
Area	40 m × 40 m
Number of nodes for patients	20
Number of nodes for doctors	10
Deployment of method	Random
Energy	0.2 J

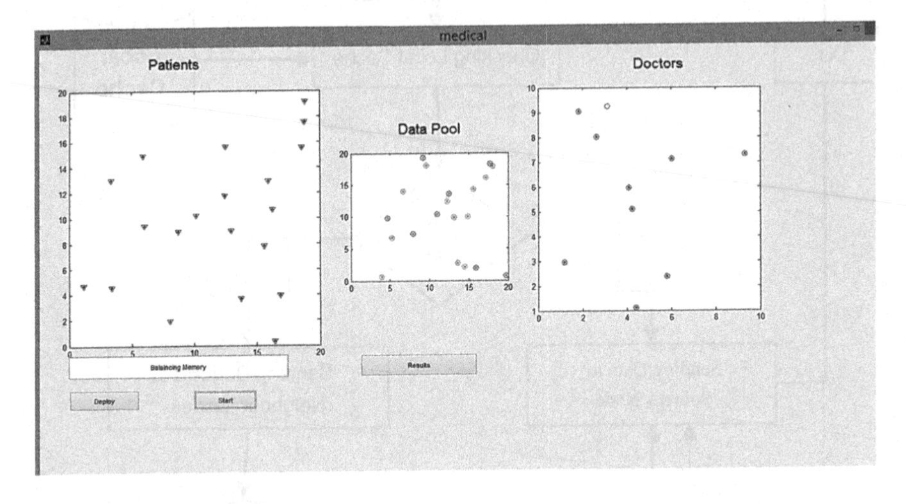

Fig. 2 Deploy sensor

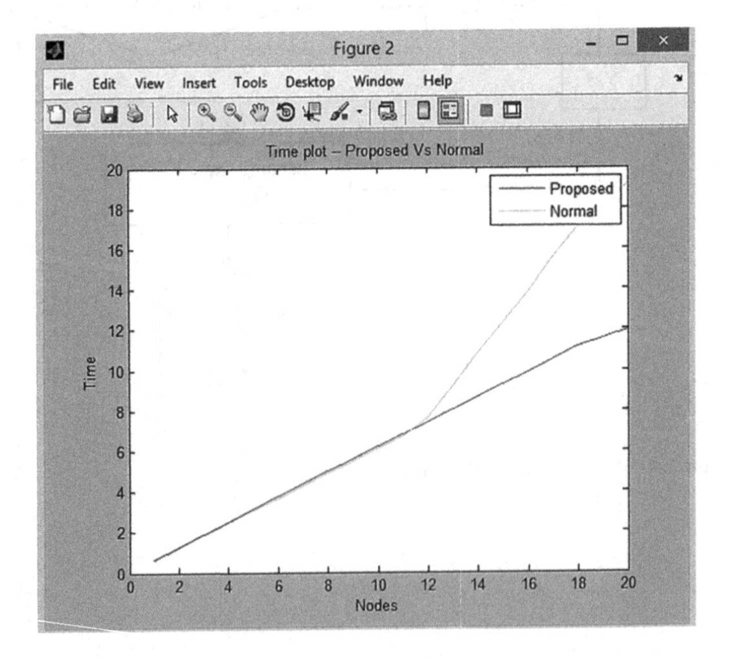

Fig. 3 Time versus node plot between proposed and normal method

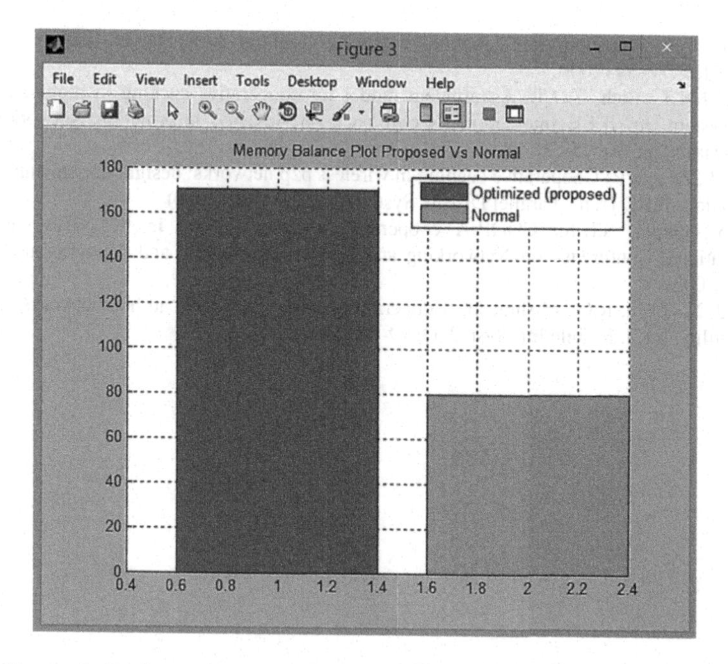

Fig. 4 Memory balancing performance between proposed method and the normal method

7 Conclusion

In this paper, we proposed SONIC scheme based on the stability of node. Firstly, the node requests data from all neighbor nodes. If the node is unable to find data, then it will send its request to data pool. After getting response, all nodes update the cache memory and the data pool memory and the desired data are replaced to that node whose stability is maximum. In this paper, SONIC scheme increases the data availability and the data access time and reduces the traffic near nodes and Data Pool.

References

1. Minh, T.N.T., Bich, T.D.T.: An efficient cache replacement strategy in mobile cooperative caching. In: IEEE International Conference on Wireless Communication Networking and Mobile Computing, pp. 1–4 (2011)
2. Chand, N., Chauhan, N., Kumar, P.: Enhancing data availability in MANETs with cooperative caching. Int. J. Mobile Comput. Multimedia Commun. **3**(4), 53–66 (2011)
3. Chiu, G.M., Young, C.R.: Exploiting in-zone broadcasts for cache sharing in mobile ad hoc networks. IEEE Trans. Mob. Comput. **8**(3), 384–397 (2009). doi:10.1109/TMC.2008.127
4. Gao, W., Srivatsa, M.: Cooperative caching for efficient data access in disruption tolerant networks. IEEE Trans. Mob. Comput. **13**(3), 611–625 (2014)

5. Yin, L., Cao, G.,: Supporting cooperative caching in ad hoc networks. In: IEEE Trans. Mob. Comput. **5**, 77–89 (2006)
6. Minh, T.N.T., Bich, T.D.T., An efficient model for cooperative caching in mobile information system. In: IEEE International Conference on Advance Information Networking and Application, pp. 90–95 (2011)
7. Zhao, J., Zhang, P.: Cooperative caching in wireless p2p networks: design, implementation and evaluation. IEEE Trans. Parallel Distrib. Syst. **21**(2), 229–241 (2010)
8. Du, Y., Gupta, S.K.S.: COOP-A cooperative caching service in MANETs. In: IEEE International Conference on Networking and Service, Autonomic and Autonomous System, p. 58 (2005)
9. Chand, N., Joshi, R.C., Mishra, M.: Cooperative caching in mobile ad hoc networks based on data utility. Int. J. Mobile Inf. Syst. **3**(1), 19–37 (2007)

Advanced Energy Sensing Techniques Implemented Through Source Number Detection for Spectrum Sensing in Cognitive Radio

Sagarika Sahoo, Tapaswini Samant, Amrit Mukherjee and Amlan Datta

Abstract The world of wireless technology is been one of the most progressive and challenging aspects for the users and providers. It deals with the wireless spectrum whose efficient use is of foremost concern. These are improved by the cognitive radio users for their noninterference communication with the licensed users. Spectrum holes detection and sensing is a dynamic time variant function which is been modified using the proposed source number detection and energy detection. Energy detection technique is implemented so as to compare the thresholds of the channels dynamically, and source detection method is used for predicting the number of channels where the energy detection is to be performed. The simulation results show the optimization and reduced probability of miss detection considering the change in threshold.

Keywords Cognitive radio · Source number detection · Energy detection · Spectrum sensing

1 Introduction

Spectrum sensing has so far been identified as the step of crucial importance in the process of the cognition cycle and the most important function for the establishment of cognitive radio network that principally emphasizes on sensing the spectrum environment accurately and determines whether the primary user is

S. Sahoo (✉) · T. Samant · A. Mukherjee · A. Datta
School of Electronics Engineering, KIIT University, Bhubaneswar, India
e-mail: sahoo09.sagarika@gmail.com

T. Samant
e-mail: itapaswini@yahoo.co.in

A. Mukherjee
e-mail: amrit1460@gmail.com

I.K. Sethi (ed.), *Computational Vision and Robotics*, Advances in Intelligent Systems and Computing 332, DOI 10.1007/978-81-322-2196-8_21

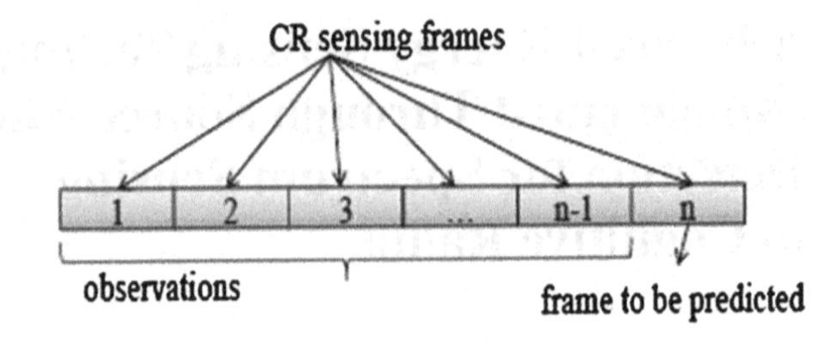

Fig. 1 Frame of sensing structure for cognitive radio

active or not over a specific band reliably [1]. Therefore, in order to guarantee noninterference with the primary user, cognitive radio must detect very weak signals [2].

In this paper, the bootstrap-based source number detection (SND) technique is applied for spectrum sensing in cognitive radio networks (CRN). A novel test source number estimation method based on bootstrap is proposed. From the simulation results, it is seen that the proposed bootstrap-based source detection procedure can provide satisfied detection performance while only requires the optimal likelihood ratio and threshold compared with the existing methods.

Using simulations, we show that when the observations of number of sources at the sensors are added on the threshold dramatically falls yielding to control in likelihood ratio to improve condition of missed detection (P_m).

It is clear from Fig. 1 that the predictor unit initially observes $n - 1$ previous sensing frames that can be used to predict the succeeding nth sensing frames.

2 Energy-based Detection

Energy detection is one of the sub-optional signal sensing techniques which has been hugely used in radio communications [3, 4]. The detection process can be performed in time domain as well as in frequency domain. Figure 1 shows the energy detection process with the hypotheses as follow:

$$H_0 : Y[n] = W[n] \quad \text{presence of signal} \tag{1}$$

$$H_t : Y[n] = X[n] + W[n] \quad \text{absence of signal} \tag{2}$$

here, $n = 1,\ldots M$; where M is the window under observation

Here, $X[n]$ denotes the sample of the target signal having certain power 'u' and $W[n]$ is noise sample that is assumed to be additive white Gaussian noise (AWGN) having zero mean and variance same as the signal power 'u' (Fig. 2).

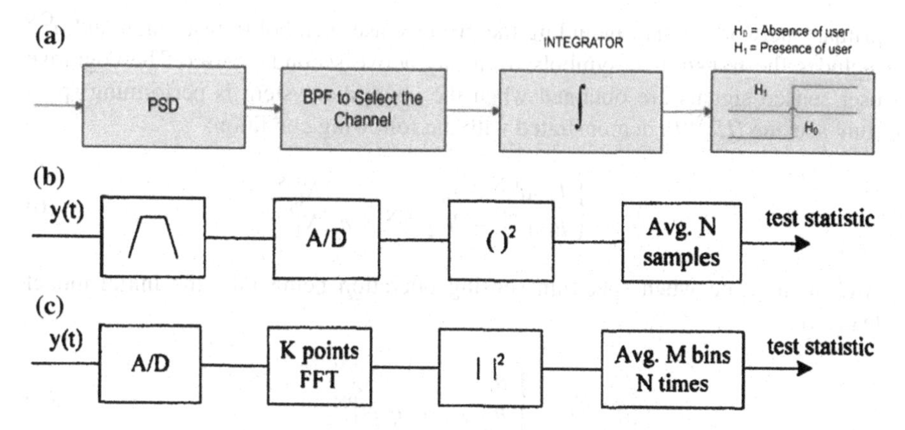

Fig. 2 The process of energy detection. **a** Spectrum sensing through energy detection. **b** Implementation in time domain. **c** Implementation in frequency domain

The sensing is performed to make certain if any activity of the primary user for a particular band of frequency occurs, as suggested by binary hypothesis testing, and that can be mapped as [5-7]:

N_0: The idle primary user

N_1: The working primary user

The presence or absence of primary signal has been given by N_1^{PN} where $i = 0$ is the absence and $i = 1$ shows the presence. Given CN i N, which is sensing and making the decision on the basis of that hypothesis modeled structures are used which is the below conditional probabilities:

$$P_m = p\left(N_0^{\mathrm{CN}} \middle| N_1^{\mathrm{PN}}\right) \tag{3}$$

$$P_{\mathrm{fa}} = P\left(N_1^{\mathrm{CN}} | N_0^{\mathrm{PN}}\right) \tag{4}$$

where P_m denotes probability of missed detection and P_{fa} as probability of false alarm. The signal which is consequently received at any particular instance say jth is given by:

$$y_j = h_{j;1}a^{\mathrm{PN}} + h_{j;2}a^{\mathrm{CN}} + n_j \tag{5}$$

and where the channel between the primary user is denoted by $h_{j;1}$ and the jth secondary user, and $h_{j;2}$ is the channel between jth secondary user and any other secondary user with setting channel property as Rayleigh distribution. a^{PN} is a primary network signal, including the primary user symbol transmission and a^{CN} includes the transmitted symbols from any active secondary user. The cognitive user sensed signals are obtained when the cognitive system is performing spectrum sensing ($H^{\mathrm{CN}}0$), demonstrated with the following condition:

$$y_j = \begin{cases} h_{j,2}a^{\mathrm{CN}} + n_j & N_0^{\mathrm{PN}} \\ h_{j,1}a^{\mathrm{PN}} + h_{j,2}a^{\mathrm{CN}} + n_j & N_1^{\mathrm{PN}} \end{cases}, \tag{6}$$

And at the time when spectrum sensing operation being idle, the initial model leads to:

$$\mathrm{y}_j = \begin{cases} n_j & N_0^{\mathrm{PN}} \\ h_{j,1^{a^{\mathrm{PN}}}} + n_j & N_1^{\mathrm{PN}} \end{cases}, \tag{7}$$

Now, the false alarm probability and detection probability can be expressed as follow:

$$\begin{aligned} Q_{d,\mathrm{SLC}} &= Q_{mK}\left(\sqrt{2\gamma_{\mathrm{slc}}}, \sqrt{\lambda}\right) \\ Q_{d,\mathrm{SLC}} &= \frac{\Gamma\left(mK, \frac{\lambda}{2}\right)}{\Gamma(mK)} \end{aligned} \tag{8}$$

where $\gamma_{\mathrm{slc}} = \sum_{k=1}^{K} \gamma_k$, y_k is the SNR received at kth cognitive user. To converse to a selection like correlation metric and/or negentropy metric, diverse measurements are utilized [8]. These are further adapted to measure the inverse gaussianity of the identified signal [9, 10]. The inverse gaussianity measured is compared with the threshold that has been presumed, as a result, the decision regarding the primary signal in the frame is within the sensing period (Fig. 3).

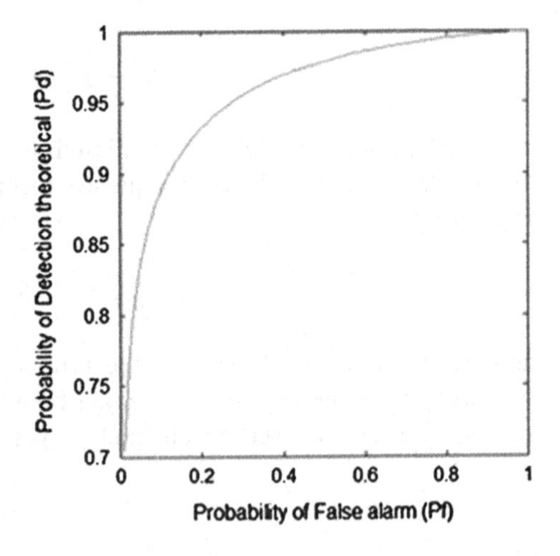

Fig. 3 Probability of detection (theoretical) versus probability of false alarm

The choice made is executed to rest of frames that are sensed. The graph shows probability of detection (theoretical) versus probability of false alarm for 1,000 simulations using the standard formula of error function.

3 Bootstrap-based Detection (Parametric and Nonparametric Resampling Method)

The common assumption of Gaussian data seems to lead to some variant of the statistic, as shown, irrespective of the source detection scheme is based on hypothesis tests or information theoretic criteria,

$$\frac{\left(\prod_{j=k}^{p} l_i\right)^{(1/(p-k+1))}}{\frac{1}{p} - k + 1 \sum_{i=1}^{p} l_i} \qquad k = 1, \ldots, p-1 \tag{9}$$

which is the ratio of the geometric mean to the arithmetic mean of the smallest sample eigenvalues [11, 12]. This can be accomplished by considering the following set of hypothesis tests for determining the number of sources:

$$\begin{aligned} H_0 &: \lambda_1 = \cdots = \lambda_p \\ &\vdots \\ H_k &: \lambda_{k+1} = \cdots = \lambda_p \\ &\vdots \\ H_{p-1} &: \lambda_{p-1} = \lambda_p \end{aligned} \tag{10}$$

with corresponding alternatives K_k, not H_k, $k = 0,\ldots,p - 2$. Acceptance of H_k leads to the estimate $q = k$. A practical procedure to estimate starts with testing and proceeds to the next hypothesis test only on rejection of the hypothesis currently being tested. Upon acceptance, the procedure stops, implying all remaining hypotheses are true. The procedure is outlined in Table 1.

Table 1 Hypothesis test procedure used for determining the number of sources

Step 1	Set $k = 0$
Step 2	Test H_k
Step 3	If H_k is accepted then set $\hat{q} = k$ and stop
Step 4	If H_k is rejected and $k < p - 1$ then set $k \leftarrow k + 1$ and return to step 2. Otherwise set $\hat{q} = p - 1$ and stop

4 Proposed Method

In the proposed method, the primary sources are been estimated through bootstrap-based SND in the initial stage and the optimization is been performed on the basis of number of sources in the later.

Figure 4 shows the detection of number of sources through an additional feedback to the traditional energy detection scheme which includes threshold $V(t)$ and the output of energy detector.

The shortcomings of the traditional spectrum sensing along with its optimization have been modified [13] which includes improper access of the spectrum efficiently and not being able to identify the presence or absence of PU [14]. To alleviate this problem, the primary sources are efficiently detected using SND technique and spectrum sensing is performed with the proposed energy-based scheme [15]. The overall model has been modified with detection of spectrum while the primary user and CR user work simultaneously.

$$\begin{aligned} &T_{i,j} = l_i - l_j, \\ &\text{where } i = k+1 \ldots p-1 \text{ and } j = i+1, \ldots, p \end{aligned} \tag{11}$$

These differences will be small when both l_i and l_j are considered to be noise eigenvalues but relatively large if any one or both of l_i and l_j are source eigenvalues. The pair wise comparisons represented in a hypothesis testing framework gives

$$H_{ij}{:}\lambda_i = \lambda_j \tag{12}$$

$$\begin{aligned} &K_{ij}{:}\lambda_i \neq \lambda_j, \\ &\text{where } i = k+1, \ldots, p-1 \text{ and } j = i+1, \ldots, p \end{aligned} \tag{13}$$

The hypotheses H_k can be reformulated as intersections between the pairwise comparisons

$$H_k = \bigcap_i^j H_{ij} \quad \text{and} \tag{14}$$

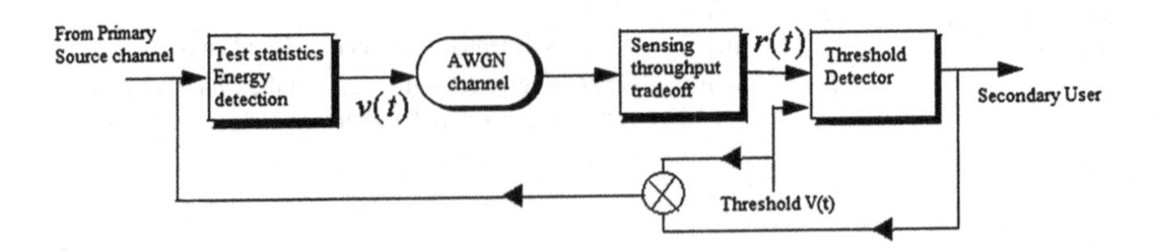

Fig. 4 Energy sensing through source number detection

$$K_k = \bigcup_i^j K_{ij}$$
$$\text{where } i = k + 1, \ldots. p - 1 \text{ and } j = i + 1, \ldots, p \tag{15}$$

One of the most popular approaches for composite hypotheses testing stands to be the generalized likelihood ratio test. By their maximum likelihood estimates, the GLRT replaces any unknown parameters. The GLRT can generally have the form

$$\hat{T}(X_N) = \frac{f_1\left(X_N; \hat{\theta}_1\right)}{f_0\left(X_N; \hat{\theta}_0\right)} \begin{cases} > \tau, \text{ accept H}_1 \\ < \tau, \text{ accept H}_0 \end{cases}, \tag{16}$$

where $\hat{\theta}_1$ is the MLE of . θ_1 assuming H_1 is true, and $\hat{\theta}_0$ is the MLE of θ_0 assuming H_0 is true.

As in the simple hypotheses, the threshold τ is found from the nominal value of the probability of false alarm P_{fa}.

5 Simulation and Discussion

The simulated results as performed is shown to using information theoretic criteria which can detect the primary users or source transmitters signal and then the spectrum sensing is carried out using bootstrap-based energy detection and

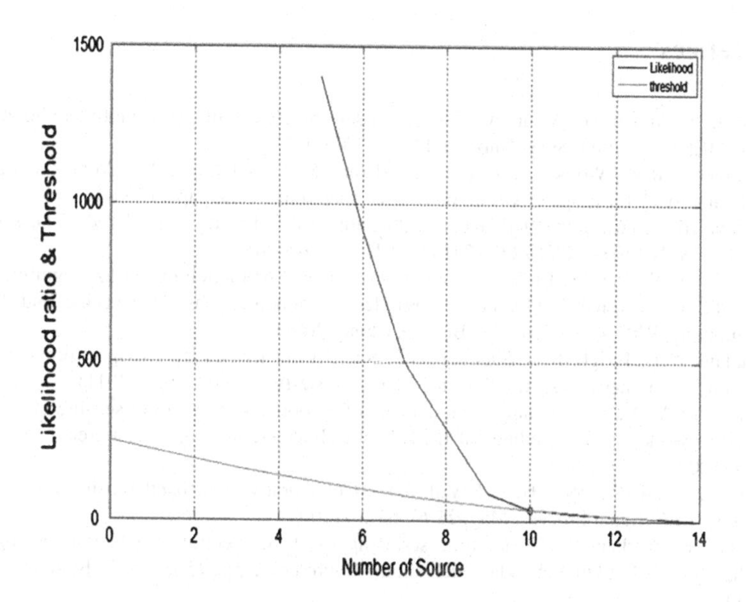

Fig. 5 Source detection for primary transmitting signals

optimization is performed as shown in Fig. 5. As shown in Fig. 5, the graph shows the number of sources versus their corresponding likelihood and threshold.

As shown in Fig. 5, the increase in number of primary sources corresponds to decrease in likelihood ratio of miss detection and also the decrease in threshold. Here, the threshold is getting reduced because of increasing number of primary sources, i.e., increase in number of primary channels implies decrease in overall selection of threshold for cognitive receiver in the time domain. There will be change in power levels for different channels as per the application and cognitive receiver is selecting the optimum threshold among all the channels. Here, the numbers of sources are calculated by comparing the likelihood and threshold of the generating signal which is 10.

6 Conclusion

In the proposed method, the likelihood of probability of miss detection is been improved with varying threshold. Here, the number of channels is considered with respect to the SND and accordingly the energy detection of individual channels is being considered. The output of the simulated graphs shows the optimum number of sources by changing the number of primary sources and the number of cognitive receivers. In practical cases, we can select the number of primary users and energy detection statistics according to the applications of different spectrum channels by cognitive users.

References

1. Yücek, T., Arslan, H.: A survey of spectrum sensing algorithms for cognitive radio applications. IEEE Commun. Surv. Tutorials **11**(1), 116–130 (2009)
2. Broderson, R.W., Wolisz, A., Cabric, D., Mishra, S.M., Willkomm, D.: Corvus: a cognitive radio approach for usage of virtual unlicensed spectrum. White paper (2004)
3. Mitola III, J.: An integrated agent architecture for software defined radio. Dissertation (2000). ISSN 1403-5286 ISRN KTH/IT/AVH—00/01—SE
4. Chunli, D., Yuning, D., Li, W.: Autoregressive channel prediction model for cognitive radio. In: 5th International Conference on Wireless Communications, Networking and Mobile Computing, WiCom '09, pp. 1–4. Beijing, China (2009)
5. Khajavi, N.T., Ivrigh, S.S., Sadough, S.M.S.: A novel framework for spectrum sensing in cognitive radio networks. IEICE Trans. Commun. **E9-B**(9), 2600–2609 (2011)
6. Ma, J., Li, Y.: Soft combination and detection for cooperative spectrum sensing in cognitive radio networks. In: Proceedings of IEEE Global Telecommunication Conference, pp. 3139–3143 (2007)
7. Quan, Z., Cui, S., Poor, H.V., Sayed, A.H.: Collaborative wideband sensing for cognitive radios. IEEE Signal Process. Mag. **25**(6), 63–70 (2008)
8. Arslan, H., Yarkan, S.: Binary time series approach to spectrum prediction for cognitive radio. In: IEEE 66th Vehicular Technology Conference, pp. 1563–1567. Baltimore, USA (2007)

9. Qiu, R.C., Chen, Z.: Prediction of channel state for cognitive radio using higher-order hidden Markov model. In: Proceedings of the IEEE SoutheastCon, pp. 276–282. Concord, USA (2010)
10. Hyvarinen, A.: Fast and robust fixed-point algorithms for independent component analysis. IEEE Trans. Neural Netw. **10**(3), 626–634 (1999)
11. Li, Y., Dong, Y.N., Zhang, H., Zhao, H.T., Shi, H.X., Zhao, X.X.: Spectrum usage prediction based on high-order markov model for cognitive radio networks. In: IEEE 10th International Conference on Computer and Information Technology (CIT), pp. 2784–2788. Bradford, UK (2010)
12. Li, Z., Shi, P., Chen, W., Yan, Y.: Square law combining double threshold energy detection in Nakagami channel. Int. J. Digit. Content Technol. Appl. **5**(12), (2011)
13. Mukherjee, A., Datta, A.: Spectrum sensing for cognitive radio using quantized data fusion and hidden Markov model. In: Proceedings of IEEE, pp. 133–137 (ISBN: 978-1-4799-2981-8/14)
14. Akbar, I., Tranter, W.: Dynamic spectrum allocation in cognitive radio using hidden Markov models: Poisson distributed case. In: Proceedings of IEEE SoutheastCon, pp. 196–20 (2007)
15. Teguig, D., Scheers, B., Le Nir, V.: Data fusion schemes for cooperative spectrum sensing in cognitive radio networks. In: Communications and Information Systems Conference, pp. 1–7 (2012). ISBN: 978-1-4673-1422-0

Dynamic Load Balancing for Video Processing System in Cloud

S. Sandhya and N.K. Cauvery

Abstract Cloud computing is one of the more desired technologies in the recent times. It provides a wide range of services to users, common being the reliable virtual environment for storage and computation. With the demand for video content/video applications increasing rapidly over the years, real-time video streaming is becoming attractive with applications such as Video on Demand (VoD) and video conferencing. Streaming applications are resulting in increased traffic; thus, load on the network is increasing. Further worsening this situation is the user demanding for higher quality of video. Video application requires more storage and bandwidth resulting in a significant load on the network, and hence, a solution combining the cloud technology with multimedia is designed for balancing load in networks.

Keywords Cloud · Video streaming · Load balancing

1 Introduction

The term 'cloud' in cloud computing [1] refers to both hardware and software to provide the aspects of computing as a service. Cloud computing provides features such as the ability to scale up and down, self-provisioning and deprovisioning of the service, and API's and billing of the service on the go. A cloud-based environment is capable of providing elastic services [2, 3]. It enables application developers to expand and

S. Sandhya (✉)
Department of Computer Science and Engineering,
R.V. College of Engineering, Bangalore, India
e-mail: sandhya.sampangi@rvce.edu.in

N.K. Cauvery
Department of Information Science and Engineering,
R.V. College of Engineering, Bangalore, India
e-mail: cauverynk@rvce.edu.in

I.K. Sethi (ed.), *Computational Vision and Robotics*, Advances in Intelligent Systems and Computing 332, DOI 10.1007/978-81-322-2196-8_22

contract the various types of services on demand. For instance, when an application requires the architecture to scale up or down to meet the demands of the clients, the cloud handles all of these by provisioning or deprovisioning of resources as required. Thus, there exists a need to balance requests between multiple clients and their instances with an appropriate mechanism. Such a mechanism is load balancing. Cloud also delivers stable and adaptive resource management mechanisms [4, 5]. Load balancing is a networking method to distribute workload across multiple nodes in a network. Load balancing [6] is employed to achieve optimal resource utilization, maximize throughput, and minimize response time by avoiding overload condition in networks.

In this paper, a mechanism that enables dynamic load balancing and resource management for video processing system is proposed. In order to handle the increase in the load due to multiple requests arriving at a given server, load balancing is to be implemented. Since video files need more storage, managing storage on servers is also an important aspect.

2 Background and Related Work

This section discusses the various approaches and mechanisms used for balancing load and streaming video. An effective load balancing algorithm [1] is described. It describes that the research challenge is to design a mechanism to spread the multimedia service task load on servers with the minimal cost for transmitting multimedia data between server clusters and clients, while the maximal limit of each server cluster is not violated.

Moghal and Mian [6] discusses prediction-based dynamic resource allocation algorithms to scale video transcoding service on a given infrastructure as a service cloud. The proposed algorithms provide mechanisms for allocation and deallocation of virtual machines to a cluster of video transcoding servers in a horizontal fashion.

Choi et al. [7] describe that the demand of video contents has rapidly increased in the past years. One of the services that have especially become attractive to the customers is real-time Video on Demand (VoD) because it offers an immediate streaming of a large variety of video contents. The price that the operators have to pay for this convenience is the increased traffic in the networks, which are becoming more congested due to the higher demand for VoD contents and the increased quality of the videos.

As a solution to the above [8], a hierarchical network system for VoD content delivery in managed networks is proposed which implements redistribution algorithm and a redirection strategy for optimal content distribution within the network core and optimal streaming to the clients. It also balances the load between the servers concentrating the traffic to the edges of the network.

Chieu [9] states that video streaming application is one of the most challenging services. Because of time constraint and variable bit rate (VBR) property of a video, the problem of streaming high-quality video which was captured and uploaded is insufficient for external WAN bandwidth. Thus, a need arises to eliminate insufficient WAN bandwidth across Internet.

3 Proposed Scheme

The paper proposes a generalized framework for balancing load on client and server. In case of virtualized cloud-based services, there might be a situation where multiple requests are sent to a given server at the same time. This increases the load on a server. To balance this load, the most frequently accessed files are duplicated onto the other buckets. This duplication distributes the load on the server. While downloading a file, if client is busy with other computations, the interface precomputes the load on the client and depending on the load a higher/ lower quality video is downloaded. The system also implements live streaming. The video is captured, compressed, and uploaded within the given time constraint. Simultaneously during capture, if another client requests to live stream the captured video, then the interface maps the time stamp of the request with the latest segment of video uploaded as .rar file and provides the latest file for streaming. Hence, with negligible delay, the video is live streamed.

The approach that delivers video service using the Internet Protocol over cloud infrastructure via a broadband connection is described. Load balancing is done at both the ends; server and client during upload and download of videos. For storage purpose, Amazon S3 buckets were used to simulate the cloud that acts as server. This provides the functions such as load balancing and resource management. It is ensured that the size of video files does not exceed the limit for a bucket and multiples requests for a bucket do not overload the server. When a bucket cannot handle the arriving user requests, it transfers the video files to another bucket to continue providing service to the user, thus balancing load at the server by distributing the user requests. At the server, when a file is requested for download, based on the available bandwidth, the quality of video provided to the user also differs (in terms of bit rate). And at the client end, each time a user uploads file, if the bucket is full then a new bucket is created and the file is uploaded thus achieving resource management. There are two types of users, namely the system admin and regular users. The admin has the privilege to view the logs which are useful in analyzing the statistics of the usage of the services offered by the framework. He can also exercise other privileged functionality such as adding or deleting users and groups, view records of users, storage and files, and granting access permission to the regular user. The regular user is one who has subscribed to use the service. The user can perform operations such as upload, download, and live stream videos.

A sequence diagram is a kind of interaction diagram that shows how processes operate with one another and also indicate their order. A sequence diagram shows object interactions arranged in time sequence. Figure 1 shows the sequence diagram illustrating the sequence of operations that are performed. When the client wants to upload a file, the client selects and sends the video file to be uploaded to the interface. The interface then gets the client's statistics and checks if he can avail the service based on the subscription. The video file is not uploaded to the current bucket either if enough space is not available or the bucket cannot handle the request. In that case, a new bucket is created and allocated to upload the file. If the client wishes to download a file, the user would select the file in the cloud

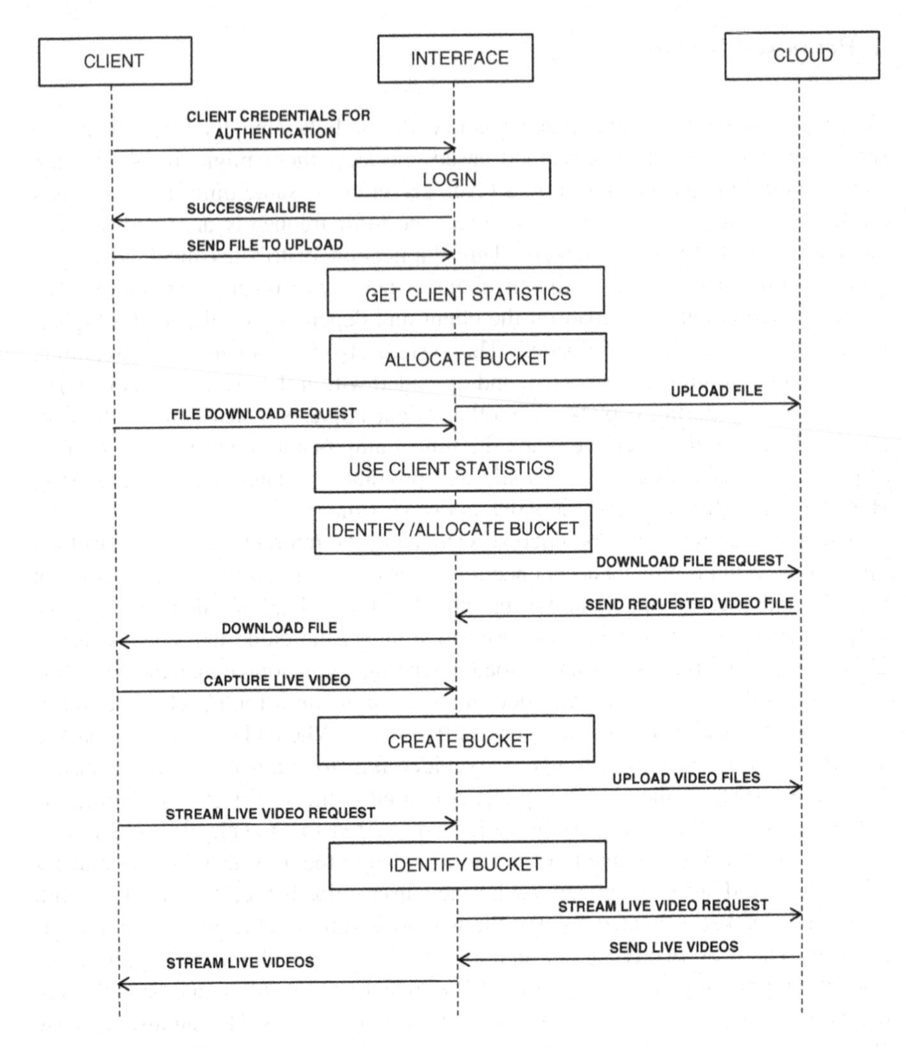

Fig. 1 Sequence diagram

and send the request to the interface. The interface then uses the client's statistics. The statistics include the information about the services previously availed by the client and the services that the user has subscribed for. The bucket from which the video has to be downloaded is identified and if it cannot handle the request, due to multiple simultaneous requests, the file is copied to another bucket which the user is provided the service. If the client end does not have the required bandwidth, then a low-resolution file of different format is given to the client that meets the bandwidth requirement.

In live video capture module, the video is captured through the camera and compressed by selecting the various video codecs. A bucket is created and the video files are uploaded into the bucket after compressing. To stream live video, the client sends a request for streaming which then downloads the videos from the bucket, uncompressed and streams it to the terminal client.

4 Methodology

Modular approach is used to implement this framework, and the following main modules are identified cloud download, cloud upload, capture, and live stream video.

4.1 Cloud Download

When the user requests for a file, it needs to be downloaded from the cloud.

The pseudo code for this module is as below:

```
Get the requested file name;
Identify and retrieve the bucket name in which the video file is stored;
verify the download status for overload condition;
        if overload then stop download;
      else
        check the bucket availability for multiple simultaneous request;
if request's serviced == threshold max
      Duplicate the bucket contents;
      Direct all new requests to this;
else
    initiate download file from the original bucket;
    determine the available bandwidth using SLoPS;
      if Bandwidth > file size to be download
              Proceed with download;
      Else
        Encode and convert the file format to low  Resolution file resulting in lower size;
          Then proceed with download;
stop
```

The interface first checks for the download status to determine whether the file can be downloaded within the specified time by checking the estimated available bandwidth, number of requests in the queue of the bucket. If the file download is estimated to exceed the time constraint, then the download is terminated. It then checks the bucket availability. The parameter threshold max stores the maximum requests that a bucket can service for uninterrupted user experience. This is determined by considering the ideal round trip time, queue lengths, and the bandwidth. If the bucket is servicing less than threshold max, then it is allowed to take the incoming request and provide service else an identical duplicate copy of the bucket is created.

This newly created bucket is referred to by a pointer that the original bucket owns. Henceforth, the requests to this bucket are forwarded to the newly created bucket for providing uninterrupted service to users. This forwarding of requests balances load at the server. Also, the available bandwidth at the client end is estimated through the technique of self-loading periodic streams (SLoPS) [10]. After which, the file size to be downloaded (in normal cases a HD mp4 file format) is compared with the estimated available bandwidth. If required bandwidth is available at client end, the HD video file is downloaded or else the file is encoded and converted to a file format with lower resolution which results in lower file size and provided to the client, thus achieving load balancing at the client end.

4.2 Cloud Upload

When the user requests for uploading a video file, the path of the selected video file to be uploaded is retrieved and the request is sent to upload along with the bucket name and filename. The bucket is allocated for video upload depending on the current status of its availability. The pseudo code for this module is as below:

```
Select a file to upload;
      If list =  empty and seqNum=0
//true if this is the first file to be uploaded to the bucket
          Assign file to bucket;
          Copy file to bucket;
          Increment seqNum;
          Stop
//Verify overload status of bucket
        else
          Get list_size;
              If list_size less than max_list
                  Add file to the list for upload;
                                  Increment seqNum;
                  Stop;
              Else
                   Get file length;
                       If file not empty
//compute waiting time for file to start writing to bucket
                         tempDataLength= data_length_list/file length;
                         temp_time=time_required;
                         temp_time= temp_time * tempDataLength;
                             if temp_time greater than 1000ms
                                time required more so use another bucket;
                             else
                                use same bucket for upload;
                                Add file to the list for upload;
                                Increment seqNum;
Stop;
```

Where seqNum keeps track of the elements in the list waiting in the queue of the bucket for uploading, list_size stores number of elements in the queue and max_list is the maximum number of elements in the queue that a bucket can store. Data_length_list stores the size of the contents of the list in terms of bits, file length stores the size of the file. Temp_time finally computes the waiting time or time the file waits in queue before starting its upload. Time_required is the time to complete the upload after upload started. It is calculated by having a listener class that tracks the current system time in milliseconds using the API System.currentTimeMillis() as soon as the first bit is transmitted and it also records the end time by tracking the upload of the last bit when end of file character is encountered. Therefore,

$$\text{Time\_required} = \text{end time recorded} - \text{start time};$$

4.3 Capture and Live Stream Video

The framework also can capture live video and share it to users, i.e., the user can live stream video being captured. The pseudo code for live video capture is as below:

```
Select camera interface to capture video;
Segment and Compress the captured  video;
Verify upload status;
        If connection available
            Check bucket availability;
                If available
                    Upload file to the same bucket;
                Else
                    Create new bucket and upload;
        Else
            Unable to upload so stop capture;
Stop;
```

On selecting the camera interface to capture the video, the video is captured which is then segmented for every interval of 5 s, compressed, and uploaded. For upload, the entire process described previously is repeated.

The pseudo code for live video streaming is as below:

```
Get request for live stream video;
Timestamp the received request;
Verify download status;
    If connection and bandwidth available
        Timestamp of the request compared with the timestamp of the last video segment uploaded;
        start service by providing the latest segment to user;
        continue providing video segments until live capture ends or request terminates;
    else
        display error indicating insufficient resources;
stop;
```

When the user makes request to live stream the video, the received request is time stamped. This is compared with the last segment of live video segment uploaded and the latest segment is provided to the user, thus starting the live stream of video. All segments of the same video is given the same IDs (identifier); therefore, once a segments is given to user, then onwards the following segments also continue to be streamed to user until either the video ends or the user terminates the request.

5 Results

The admin uses the logs and report for billing of service usage in pay-as-you-go model. S3 buckets stores objects up to 5 TB in size together with 2 KB of metadata. Maximum size permitted for each object is 5 GB. Hence, these objects provide means of storage management. Replication of the most frequently accessed objects in a given bucket helps to distribute and balance load when the server begins to get swamped. While converting the file format in case of client bandwidth being a constraint, a lower resolution to higher resolution (hd, mp4) takes 3 times real time, and on the contrary, converting from higher resolution to lower resolution needs 1 time real time. The speed of en/decoding has impact on perceived quality of the video. Latency is directly proportional to quality of the sender's camera. Sender with higher resolution camera needs more processing time than a low-resolution one.

6 Conclusion

The proposed framework integrates the required functionalities for balancing load on client and server. It also includes mechanism to manage storage of videos. If case of server gets swamped from multiple simultaneous requests from various clients, the load on the server is balanced by duplicating the bucket to which the new requests are directed. On the counterpart, the interface determines the load on the client for every download and depending on the status, a higher or lower resolution video is provided thus managing load at client end. The proposed work can be extended to include encryption for the video content on the cloud to provide security to videos stored on the cloud.

Acknowledgment I express my heartfelt gratitude and acknowledge the efforts of my students Krishna S Raghavan, Priya S, Shruthi R, and Sumana P.

References

1. Lin, C.C., Chin, H.H., Deng, D.J.: Members, IEEE, Dynamic Multi-service load balancing in cloud-based multimedia system. IEEE Syst. J. **8**(1) 225–234 (2014)
2. Yu, L., Thain, D.: Resource management for elastic cloud workflows. In: 12th IEEE/ACM International Symposium on Cluster, Cloud and Grid Computing (CCGrid), May 2012
3. Zhu, Z., Li, S., Chen, X.: Design QoS-Aware Multi-Path Provisioning Strategies for Efficient Cloud-Assisted SVC Video Streaming to Heterogeneous Clients. IEEE (2010)
4. Urgaonkar, R., Kozat, U., Igarashi, K., Neely, M.J.: Dynamic resource allocation and power management in virtualized data centers. In: Proceedings of IEEE IFIP NOMS (2010)
5. Jokhio, F.: Prediction-based dynamic resource allocation for video transcoding in cloud computing. In: 21st Euromicro International Conference on Parallel, Distributed and Network-Based Processing (PDP), 27 Feb 2013
6. Moghal, M.R., Mian, M.S.: Effective Load Balancing in Distributed Video-On-Demand Multimedia System. IEEE (2003)
7. Choi, J., Yoo, M., Mukherjee, B.: Efficient Video-on-Demand Streaming for Broadband Access Networks. IEEE (2010)
8. Sedano, I., Kihl, M., Brunnström, K., Aurelius, A.: Evaluation of Video Quality Metrics on Transmission Distortions in H.264 Coded Video. IEEE (2010)
9. Chieu, T.C: Dynamic scaling of web applications in a virtualized cloud computing environment, In: IEEE International Conference on e-Business Engineering (2009)
10. Hu, N.: Network Monitoring and Diagnosis Based on Available Bandwidth Measurement, CMU-CS-06-122. Carnegie Mellon University, Pittsburgh, pp. 1–51 (2006)

CT and MR Images Fusion Based on Stationary Wavelet Transform by Modulus Maxima

Om Prakash and Ashish Khare

Abstract In medical imaging, combining relevant information from the images of computed tomography (CT) and magnetic resonance imaging (MRI) is a challenging task. MR image carries soft tissue information that shows presence like tumor and CT image shows bone structures. For applications such as bioscopy planning and radio therapy, both kind of information is needed. This makes fusion problem more interesting and challenging. In this paper, we present an image fusion method based on stationary wavelet transform that decomposes source images into approximation, horizontal, vertical, and diagonal components. Coefficients of each of these components are combined using absolute maximum selection criteria separately. Inverse transformation results in a fused image. Also, the proposed method fuses images in presence of noise accurately. The performance of the proposed method is assessed visually and quantitatively. Entropy, fusion factor, and standard deviation are used as fusion performance measures.

Keywords Image fusion · Medical imaging · Stationary wavelet transform · Fusion rule · Multimodal images

1 Introduction

In medical sciences, multiple modality images are used for diagnosis of various types of diseases in human body parts. For example, X-ray images are used to find fractures in bone, and computed tomography (CT) image shows the clear

O. Prakash (✉)
Centre of Computer Education, Institute of Professional Studies,
University of Allahabad, Allahabad, India
e-mail: au.omprakash@gmail.com

A. Khare
Department of Electronics and Communication, University of Allahabad,
Allahabad, India
e-mail: ashishkhare@hotmail.com

I.K. Sethi (ed.), *Computational Vision and Robotics*, Advances in Intelligent Systems and Computing 332, DOI 10.1007/978-81-322-2196-8_23

structure of denser tissues such as bones, while magnetic resonance image (MRI) gives better pattern of soft tissues, such as brain. For diagnosis purpose, it is better if information related to soft tissues as well as hard tissues is available in a single image. Combining the information from multimodal images to produce a single composite image without introducing any artifacts is a challenging problem in computer vision [1]. In medical imaging, radiology [2], molecular and brain imaging [3], oncology [4], cardiac diseases [5], neuroradiology [6], etc., are some of the important areas in which image fusion applications are found. CT and MR images are widely used for clinical purpose in medical imaging.

Various spatial domain- [7–9] and wavelet transform domain [10, 11]-based image fusion methods have been proposed in the past. Some of the spatial domain methods are linear fusion [7], principal component analysis (PCA) [8], and sharpness criteria [9]-based fusion. However, spatial domain-based image fusion techniques often produce poor results because they usually produce spectral distortions in the fused image. In recent years, wavelet transform-based image fusion methods getting popularity due to their multiresolution decomposition ability.

There are two basic requirements for image fusion [12]. First, fused image should possess all possible relevant information contained in the source images; second, fusion process should not introduce any artifact, noise, or unexpected feature in the fused image. Most wavelet transform suffers from the shift variance and poor directionality [13]. A transform is shift variant if a change in the input signal causes an unpredictable change in transform coefficients. This unpredictable change creates erroneous fusion results [15]. Erroneous fusion may lead to wrong diagnosis in case of medical imaging. Therefore, a shift invariant stationary wavelet transform has been used for image fusion.

In this paper, we proposed a new image fusion method based on stationary wavelet transform using modulus maxima. In the proposed method, two main steps have to be followed: First, the source images are decomposed into low-pass and high-pass sub-bands of different scale using stationary wavelet transform, and second, low-pass sub-band is divided into a set low-pass and high-pass sub-bands and so on.

The rest of paper is organized as follows: Overview of stationary wavelet transform is given in Sect. 2. The proposed method and experimental results are given in Sect. 3 and 4, respectively. At last, in Sect. 5, conclusions of the work have been presented.

2 The Stationary Wavelet Transform

Stationary wavelet transform (SWT) provides efficient numerical solutions in the image processing applications. Unlike the classical discrete wavelet transform (DWT), SWT gives a better approximation because of its linear, redundant, and

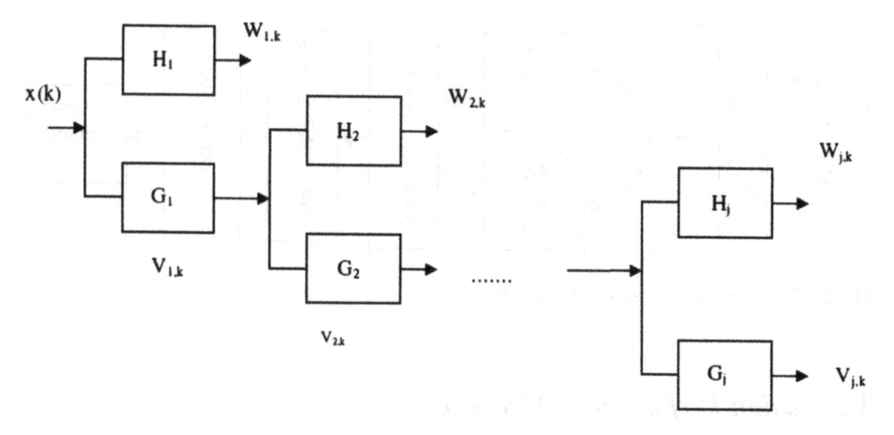

Fig. 1 Decomposition in stationary wavelet transform

shift invariance nature. A brief description of SWT is presented here. Computation of SWT for a 1-D signal $x(k)$ is shown in Fig. 1. $W_{j,k}$ and $V_{j,k}$ are called the detailed and approximation coefficients of SWT of the signal $x(k)$. The filters H_j and G_j are the standard low-pass and high-pass wavelet filters, respectively. In the first step, the filters H_1 and G_1 are obtained by upsampling the filters using the previous step (i.e., H_{j-1} and G_{j-1}) [14].

3 The Proposed Method

The proposed fusion scheme is based on the computation of stationary wavelet transform of source images. Both CT and MR images are separately transformed using SWT, and then, the coefficients obtained are fused using modulus maxima fusion rule. The overall working of fusion process is shown in Fig. 2.

3.1 Fusion Rule

In this paper, we have used modulus maxima selection fusion rule for combining SWT coefficients of pair of images. Let $M_1(x, y)$ and $M_2(x, y)$ are the two medical images to be fused and their stationary wavelet coefficients are $\mathrm{SWT}_1(m, n)$ and $\mathrm{SWT}_2(m, n)$, respectively, then, modulus maxima selection fusion rule combines stationary wavelet coefficients as

$$W(m,n) = \begin{cases} \mathrm{SWT}_1(m,n), & \text{if}\, \mathrm{SWT}_1(m,n) \geq \mathrm{SWT}_2(m,n) \\ \mathrm{SWT}_2(m,n), & \text{if}\, \mathrm{SWT}_2(m,n) > \mathrm{SWT}_1(m,n) \end{cases}$$

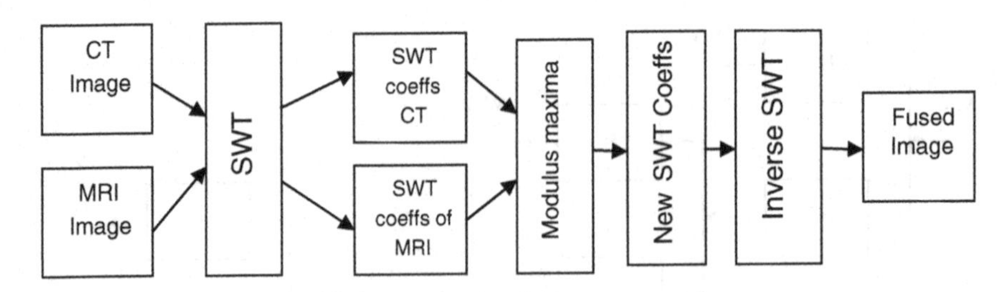

Fig. 2 The proposed image fusion method

3.2 Fusion Performance Measures

(i) **Standard Deviation** (σ)

Standard deviation is the measure of the contrast of the fused image and it can be calculated as

$$\sigma = \sqrt{\sum_{i=0}^{L-1} \left(i - \bar{i}\right) h_F(i)}, \quad \bar{i} = \sum_{i=0}^{L-1} i h_F(i)$$

where i, h_F, and L are the gray-level index, the normalized histogram of the fused image, and the number of bins in histogram, respectively [15]. Higher the value of standard deviation, better is the quality fused image.

(ii) **Entropy**

The entropy is the amount of information contained in fused image. It is calculated as

$$Q = -\sum_{i=0}^{L-1} P_i \log_2 P_i$$

where L is the number of gray level and P_i is the ratio between the number of pixels with gray values i and total number of pixels.

(iii) **Fusion Factor (FF)**

It is the sum of the mutual information [15] of source images and fused image

$$FF = M_{AF} + M_{BF}$$

where M_{AF} and M_{BF} are mutual information between source images and fused image.

4 Experimental Results

The fusion of CT and MR medical images has been performed using MATLAB. The experiments are performed on different CT and MR image pairs. For representation purpose, fusion results of one pair of CT and MR images are shown in Fig. 3.

Robustness of the proposed method is tested by fusing the noisy images. For this, we manually made source images corrupted by zero mean Gaussian noise and fusion has been performed. The result of fusion of noisy image is shown in Fig. 4.

The results are also evaluated quantitatively using standard deviation (σ), entropy (Q), and fusion factor (FF) as fusion metrics. These metric values are shown in Table 1.

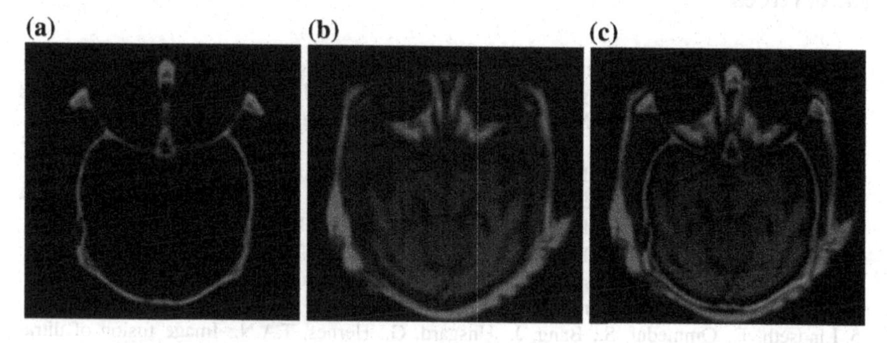

Fig. 3 Fusion result when images are free from any noise **a** CT image, **b** MR image, and **c** fused image

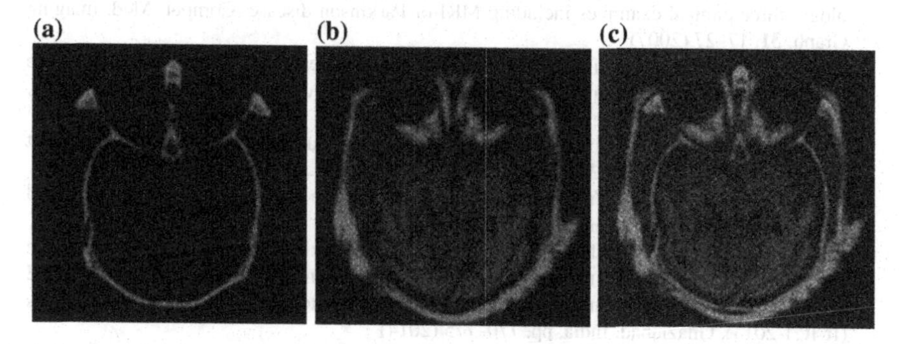

Fig. 4 Fusion result when images are corrupted by zero mean Gaussian noise of variance 0.01 **a** noisy CT image **b** noisy MR image, and **c** fused image

Table 1 Image fusion performance metric values

Fusion metric	Noise-free fused image	Noisy fused image
Standard deviation	32.9014	36.7091
Entropy	5.9902	6.5642
Fusion factor	1.3890	1.2826

5 Conclusions

This paper shows that the stationary wavelet transform can be employed in applications such as image fusion. Fusion of CT and MR images has been performed when images are free from any noise and when images are corrupted by zero mean Gaussian noise of variance 0.01. Shift invariance property of SWT retained the finer detail as well as edge information of the source images in the fused image. Fused images are suitable for better diagnosis, treatment, and human vision. These experimental results show that the proposed method performs well in both the cases.

References

1. Forsyth, D.A., Ponce, J.: Computer Vision—A Modern Approach. PHI Pvt. Ltd, India (2009)
2. Nakamoto, Y., Tamai, K., Saga, T., Higashi, Hara, T., Suga, T., Koyama, Togashi, K.: Clinical value of image fusion from MR and PET in patients with head and neck Cancer. Mol. Imag. Biol. **11**, 46–53 (2009)
3. Giesel, F.L., Mehndiratta, A., Locklin, V., McAuliffe, M.J., White, S., Choyke, P.L., Knopp, M.V., Wood, B.J., Haberkorn, U., Tengg-Kobligk, H.V.: Image fusion using CT, MRI and PET for treatment planning, navigation and follow up in percutaneous RFA. Exp. Oncol. **31**, 106–114 (2009)
4. Kannathal, N., Acharya, U.R., Ng, E.Y.K., Krishnan, S.M., Min, L.C., Laxminarayan, S.: Cardiac health diagnosis using data fusion of cardiovascular and haemodynamic signals. Comput. Methods Prog. Biomed. **82**, 87–96 (2006)
5. Lindseth, F., Ommedal, S., Bang, J., Unsgard, G., Hernes, T.A.N.: Image fusion of ultrasound and MRI as an aid for assessing anatomical shifts and improving overview and interpretation in ultrasound-guided neurosurgery. Int. Congr. Ser. **1230**, 254–260 (2001)
6. Rojas, G.M., Raff, U., Quintana, J.C., Huete, I., Hutchinson, M.: Image fusion in neuroradiology: three clinical examples including MRI of Parkinson disease. Comput. Med. Imaging Graph. **31**, 17–27 (2007)
7. Clevers, J.G.P.W., Zurita-Milla, R.: Multisensor and multiresolution image fusion using the linear mixing model. In: Stathaki, T. (eds.) Image Fusion: Algorithms and Applications, pp. 67–84. Academic Press, Elsevier, New York (2008)
8. Naidu, V.P.S., Raol, J.R.: Pixel-level image fusion using wavelets and principal component analysis. Defence Sci. J. **58**(3), 338–352 (2008)
9. Tian, J., Chen, L., Ma, L., Yu, W.: Multi-focus image fusion using a bilateral gradient-based sharpness criterion. Opt. Commun. **284**, 80–87 (2011)
10. Prakash, O., Kumar, A., Khare, A.: Pixel-level image fusion scheme based on steerable pyramid wavelet transform using absolute maximum fusion rule. In: Proceedings of IEEE International Conference on Issues and Challenges in Intelligent Computing Techniques (ICICT-2014), Ghaziabad, India, pp. 770–775 (2014)
11. Singh, R., Srivastava, R., Prakash, O., Khare, A.: Multimodal medical image fusion in dual tree complex wavelet domain using maximum and average fusion rules. J Med. Imaging Health Inf. **2**, 168–173 (2012)
12. Rockinger, O., Fechner, T.: Pixel level fusion: the case of image sequences In: Signal Processing, Sensor Fusion, and Target Tracking (SPIE), vol. 3374, pp. 378–388 (1998)
13. Zhang, Z., Blum, R.S.: A categorization of multiscale decomposition-based image fusion scheme with a performance study of digital camera application. Proc. IEEE **87**, 1315–1326 (1999)
14. Nason, G.P., Silver, B.W.: The stationary wavelet transform and some statistical applications. Technical Report BS8 1Tw, University of Bristol (1995)
15. Deshmukh, M., Bhosale, U.: Image fusion and image quality assessment of fused images. Int. J. Image Process **4**, 484–508 (2010)

A Real-Time Intelligent Alarm System on Driver Fatique Based on Video Sequence

Shamal S. Bulbule and Suvarna Nandyal

Abstract The proposed work deals with development of an automatic system for drowsy driver detection using machine vision system. The system uses a small monochrome security camera that points directly toward the driver's face and focuses the driver's eyes. The video samples of drivers in drowsy and non-drowsy condition are obtained and stored in database. The video samples are converted to frames. Each frame is converted to binary images to track edges of eyes. An algorithm is developed to locate the eyes and its closure. After extracting the face area, the eyes are located by computing the average of pixels in horizontal area. Taking into account the knowledge that eye regions in the face present great intensity changes and the eyes are located by finding the significant intensity changes in the face. Once the eyes are located, measuring the distances between the intensity changes in the eye area determined whether the eyes are open or closed. The variation in intensity is plotted. Based on the distance between two valleys of the plot eyes, closure is detected. Once the closure is detected, fatigue is reported through a warning signal to the driver to it. The algorithm developed is unique to any currently published papers, which was a primary objective of our work. The performance of the work is reported for drowsy and non-drowsy driver's samples in different environment.

Keywords Driver · Binarization · Edge · Drowsy detection · Intensity change

S.S. Bulbule (✉) · S. Nandyal
Department Computer Science and Engineering, PDA, College of Engineering, Gulbarga, Karnataka, India
e-mail: shamal.s.b@gmail.com

S. Nandyal
e-mail: suvarna_nandyal@yahoo.com

I.K. Sethi (ed.), *Computational Vision and Robotics*, Advances in Intelligent Systems and Computing 332, DOI 10.1007/978-81-322-2196-8_24

1 Introduction

Driving is a complex task where the driver is responsible of watching the road, taking the correct decisions on time, and finally, responding to other drivers' actions and different road conditions. Vigilance is the state of wakefulness and ability to effectively respond to external stimuli. Driving for a long period of time causes excessive fatigue and tiredness which in turn makes the driver sleepy or loose awareness. The study states that the cause of an accident falls into one of the following main categories: (1) human, (2) vehicular, and (3) environmental. The three main categories (human, vehicular, and environmental) are related among each other, and human error can be caused by improper vehicle or highway design characteristics. The recognized three major types of errors within the human error category are as follows: (1) recognition, (2) decision, and (3) performance. Decision errors refer to those that occur as a result of a driver's improper course of action or failure to take action. A recognition error may occur if the driver does not properly perceive or comprehend a situation. To perform all these activities in time and accurately, it is necessary that driver must be vigilant. The aim of this project is to develop an automatic system for drowsiness detection system. The focus will be placed on designing a system that will accurately monitor the open/closed state of the driver's eyes in real time. By monitoring the eyes, it is believed that the symptoms of driver fatigue can be detected early to avoid a car accident.

To study the state-of-art technology, the literature survey is carried out related to the present problem. Mr. Susanta Podder and Mrs. Sunita Roy (2013) have developed such a system we need to install some hardware components like camera inside the car, which can capture the image of the driver at a fixed interval and an alarm system, which will alert the driver after detecting his/her level of drowsiness. D. Jayanthi and M. Bommy (2012) have addressed the problem, and we propose a vision-based real-time driver fatigue detection system based on eye tracking, which is an active safety system. Eye tracking is one of the key technologies for future driver assistance systems since human eyes contain much information about the driver's condition such as gaze, attention level, and fatigue level. Nidhi Sharma and V.K. Banga (2010) have presented driving support systems, such as car navigation systems, are getting common, and they support drivers in several aspects. It is important for driving support systems to detect status of driver's consciousness.

2 Proposed Methodology

The propose methodology of the work is shown in Fig. 1. The input processes a video clip taken from the database; the video samples are converted to frames. In our work, we have taken 20 frames/images per sample.

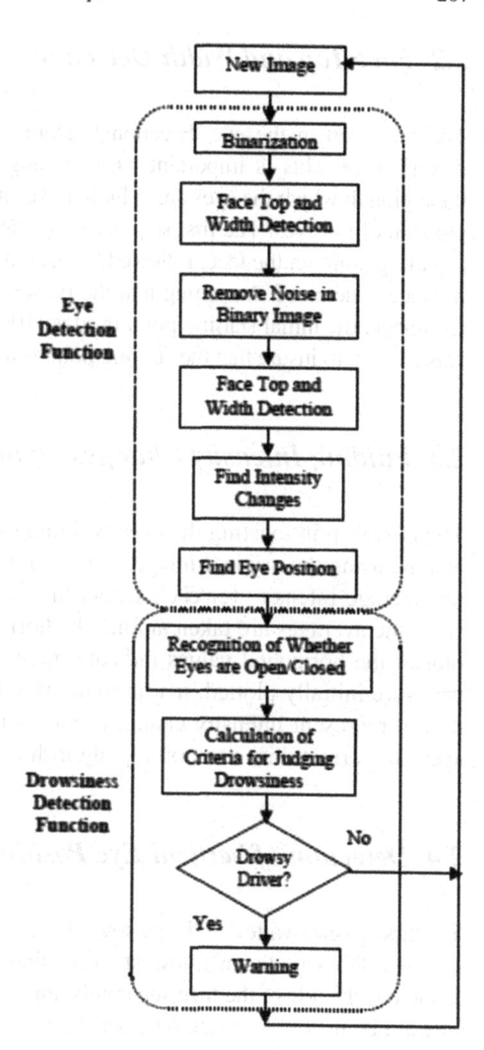

Fig. 1 Proposed work

2.1 Eye Detection

Each frame represents the facial image depicting the eye closure motion. These images are given as input successively. The preprocessing is performed by binarizing the image. The top and sides of the face are detected to narrow down the area of where the eyes exist. Using the sides of the face, the center of the face is found, which will be used as a reference when comparing the left and right eyes. Moving down from the top of the face, horizontal averages (average intensity value for each y-coordinate) of the face area are calculated. Large changes in the averages are used to define the eye area. The following explains the eye detection procedure in the order of the processing operations.

2.2 Face Top and Width Detection

The next step in the eye detection function is determining the top and side of the driver's face. This is important since finding the outline of the face narrows down the region in which the eyes are, which makes it easier (computationally) to localize the position of the eyes. The first step is to find the top of the face. The next step is to find a starting point on the face, followed by decrementing the *y*-coordinates until the top of the face is detected. Assuming that the person's face is approximately in the center of the image, the initial starting point used is (100, 240). The starting *x*-coordinate of 100 was chosen, to insure that the starting point is a black pixel (no. on the face).

2.3 Finding Intensity Changes on the Face

The next step in locating the eyes is finding the intensity changes on the face. This is done using the original image, *not* the binary image. The first step is to calculate the average intensity for each *y*-coordinate. This is called the horizontal average, since the averages are taken among the horizontal values. The valleys (dips) in the plot of the horizontal values indicate intensity changes. When the horizontal values were initially plotted, it was found that there were many small valleys, which do not represent intensity changes, but result from small differences in the averages. To correct this, a smoothing algorithm was implemented.

2.4 Detection of Vertical Eye Position

The first largest valley with the lowest *y*-coordinate is the eyebrow, and the second largest valley with the next lowest *y*-coordinate is the eye. This process is done for the left and right side of the face separately, and then, found eye areas of the left and right side are compared to check whether the eyes are found correctly. Calculating the left side means taking the averages from the left edge to the center of the face and similarly for the right side of the face. The reason for doing the two sides separately is because when the driver's head is tilted, the horizontal averages are not accurate. For example, if the head is tilted to the right, the horizontal average of the eyebrow area will be of the left eyebrow and possibly the right hand side of the forehead.

2.5 Drowsiness Detection

The state of the eyes (whether it is open or closed) is determined by distance between the first two intensity changes found in the above step. When the eyes are closed, the distance between the *y*-coordinates of the intensity changes is larger if compared to when the eyes are open. This is shown in Fig. 2. The limitation to this is if the driver moves their face closer to or further from the camera.

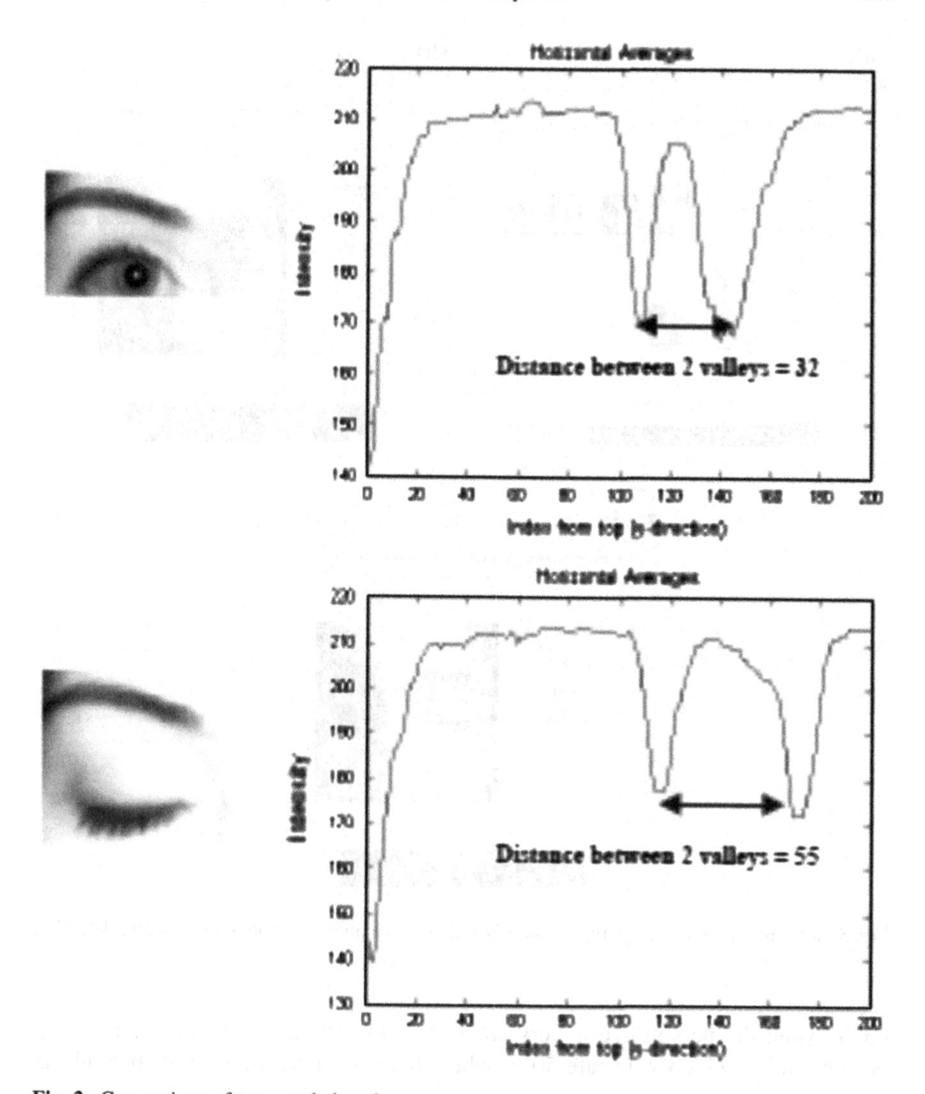

Fig. 2 Comparison of open and closed eye

If this occurs, the distances will vary, since the number of pixels the face takes up varies. Because of this limitation, the system developed assumes that the driver's face stays almost the same distance from the camera at all times.

3 Results and Discussion

The bar graph describes the output for different users, the *v*1 video with 80 % of rate of fatique describes for the person with active driving, *v*2 video with rate 70 % shows the output when the person is wearing eyeglasses, *v*3 video with rate

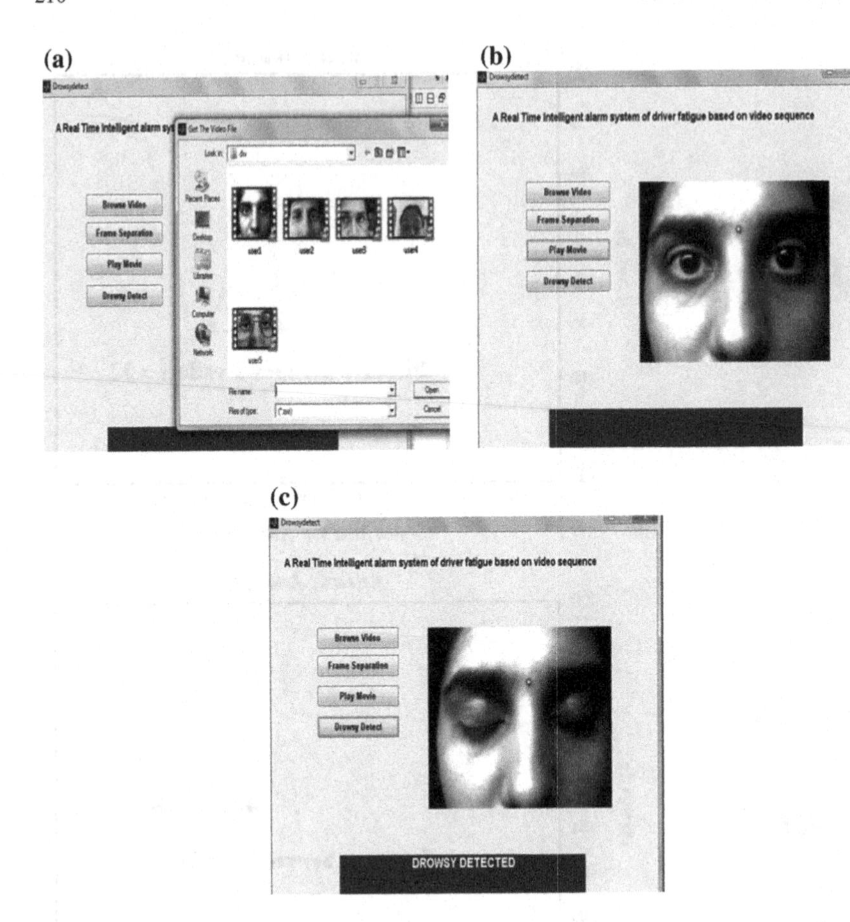

Fig. 3 **a** Input screen to accept input video. **b** Video at eye open position. **c** Drowsiness detection

65 % when the person eye is very small, *v*4 video with rate 60 % for the dark person, and *v*5 video with rate 50 % when there is a reflecting object behind the driver (Fig. 3).

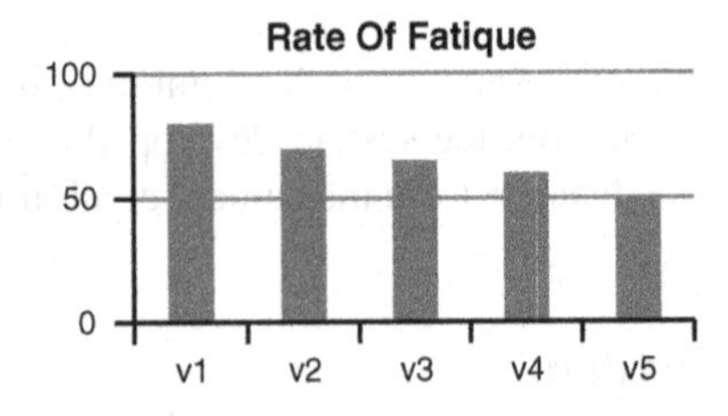

4 Conclusion and Future Scope

A noninvasive system to localize the eyes and monitor fatigue was developed. Information about the head and eyes position is obtained through various self-developed image processing algorithms. During the monitoring, the system is able to decide if the eyes are opened or closed. When the eyes have been closed for too long, a warning signal is issued. The system has been tested on 5 different subjects belonging to different races, gender, and having different skin color and facial hair. The system gave very good results for almost all of the cases. The system rarely loses track of eyes for relatively low head displacements. The system can also detect the blinks of the driver and gave no false alarms in 4 of the 5 cases studied.

Using 3D images is another possibility in finding the eyes, and this may be a more robust way of localizing the eyes. Adaptive binarization is an addition that can help make the system more robust. This may also eliminate the need for the noise removal function, cutting down the computations needed to find the eyes. This will also allow adaptability to changes in ambient light. The system does not work for dark skinned individuals. This can be corrected by having an adaptive light source. The adaptive light source would measure the amount of light being reflected back. If little light is being reflected, the intensity of the light is increased. Darker skinned individual need much more light, so that when the binary image is constructed, the face is white, and the background is black. By using these measures, the proposed work can be further improved.

References

1. Davies, E.R.: Machine Vision: Theory, Algorithms, and Practicalities. Academic Press, San Diego (1997)
2. Dirt Cheap Frame Grabber (DCFG) documentation, file dcfg.tar.z. Available from http://cis.nmclites.edu/ftp/electronics/cookbook/video/
3. Eriksson, M., Papanikolopoulos, N.P.: Eye-tracking for detection of driver fatigue. In: Proceedings of IEEE Intelligent Transport System, pp 314–319 (1997)
4. Gonzalez, R.C., Woods, R.E.: Digital Image Processing. Prentice Hall, Upper Saddle River (2002)
5. Grace, R., et al.: Drowsy driver detection system for heavy vehicles. In: Proceedings of 17th DASC Digital Avionic Systems Conference, The AIAA/IEEE/SAE, pp. I36/1–I36/8 (1998)
6. Ueno, H., Kaneda, M., Tsukino, M.: Development of drowsiness detection system. In: Vehicle Navigation Information Systems Conference, Yokohama, Japan, pp. 15–20 (1994)
7. Feraric, J., Kopf, M., Onken, R.: Statistical versus neural bet approach for driver behavior description and adaptive warning. In: Proceedings of 11th European Annual Manual, pp. 429–436 (1992)

An Efficient Android-Based Body Area Network System for Monitoring and Evaluation of Patients in Medical Care

Lakshmana Phaneendra Maguluri, G.N.B. Markandeya and G.V.S.N.R.V. Prasad

Abstract Body area network (BAN) is a promising technology for monitoring different physiological parameters of the patients in real time. Particularly, when the BAN integrated with wireless technologies provides the complete medical infrastructure. Android is the most popular operating system in the smartphone. The wireless BAN combined with an Android-based smartphone offers a large functionality. Various medical parameters can be analyzed, stored, and visualized using the GUI of an Android application designed for the end user. The different sensors placed on the body of the patient acquire physiological data from patient. This acquired data is then, gone through signal processing and data analysis and results are sent to coordinator node. The physiological data are transferred, via bluetooth to an Android-based smartphone. The physiological parameters of the patients are continuously monitored by the system and if any variation occurred, it sends alert messages to the doctor. The alert is of two types, SMS alert and email alert. Using this alert system, the emergency situation can be handled effectively and the patient will get the medical care as soon as possible.

Keywords Body area network (BAN) · Android · Smartphone · Alert

1 Introduction

The population in the world is increasing day by day. As the mortality rate reduced due to developed healthcare technologies and facilities the proportion of senior citizens is increased in the society. These senior citizens are easily vulnerable to chronic diseases, which require proper medical care than the rest of the population [1].

L.P. Maguluri (✉) · G.N.B. Markandeya · G.V.S.N.R.V. Prasad
Department of Computer Science and Engineering, Gudlavalleru Engineering College, Gudlavalleru, Andhra Pradesh, India
e-mail: phanendra51@gmail.com

I.K. Sethi (ed.), *Computational Vision and Robotics*, Advances in Intelligent Systems and Computing 332, DOI 10.1007/978-81-322-2196-8_25

So, the monitoring and recording of physiological parameters of patients outside the clinical environment are becoming increasingly important in order to take care of senior citizens.

Today, the networking technologies are very much developed. Wireless communications technologies have greatly affected on the people lifestyle. According to the report of the World Health Organization (WHO), about 17 million people die around the world due to cardiovascular diseases, particularly heart attack. Most of the deaths in them are due to untimely intervention. If proper medical care is provided to the patients at the right time, their lives can be saved [2]. Telemedicine is the field of online monitoring and analysis of vital parameters of the patients, and in the emergency situation, it helps to provide medical care as early as possible, so that the life of the patient can be saved. This telemedical field becomes very useful for the cardiac patients. So, the patient's physiological activity is continuously monitored by the system and if any variation found in the physiological activity of the patient, it informs the medical professional.

2 Body Area Network

Body area network (BAN) or body sensor network is the wireless network consists of several body sensor units connected to the central unit that performs the signal processing and data analysis. Basically, BAN consists of the BAN architecture and the communication protocol such as Bluetooth or Zigbee.

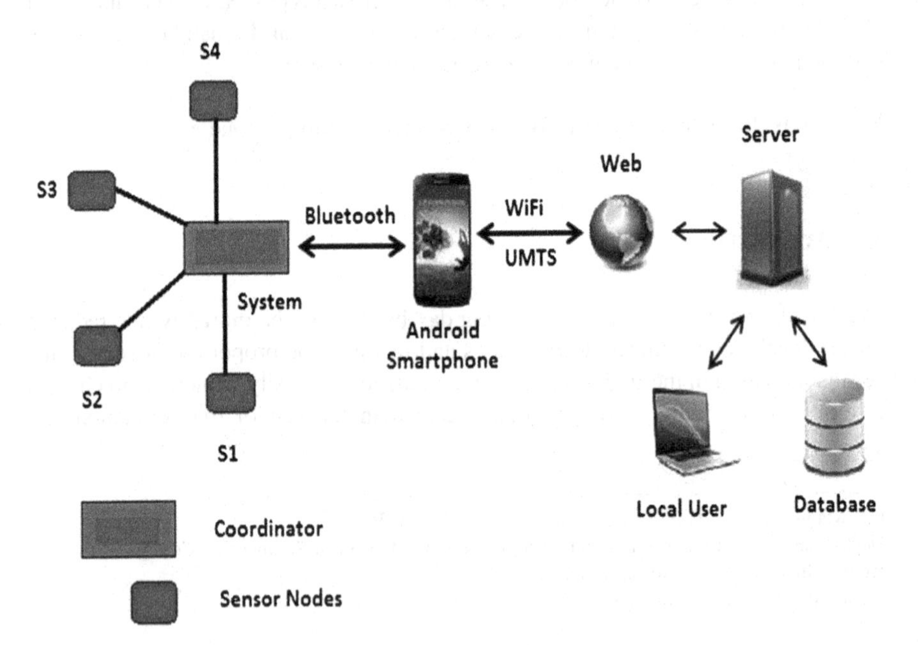

Fig. 1 BAN architecture

Wireless BAN architecture is shown in Fig. 1. In this architecture, the different sensor nodes performed the primary data processing, which includes the physiological signal processing in the microcontroller. The Android smartphone performed the secondary data processing and the resulting output can be viewed on the graphical user interface (GUI) designed for the end user. This secondary data processing includes data filtering, data representation, graphical interface, and data synchronization. This medical server allows the secure local and remote access for medical professional using Internet. In this design, Bluetooth is used to link the different sensors with the Android smartphone.

3 Electrical Components

Atmel microcontroller is used for reliable communication with the smartphone, as they are power efficient and provides an ideal platform for Tableware services [3]. MAX232 dual drivers/receivers IC basically converts the voltage level from RS-232 to the voltage level compatible with TTL logic gates. It converts only the four signals, namely RX, TX, CTS, and RTS. It operates from a single 5 V power supply.

LM353 dual operational amplifier that was chosen has the following characteristic.

- Internally compensated input offset voltage 10 mV
- Low input bias current 50 pA
- Wide gain bandwidth 4 MHz
- High slew rate 13 V/μs
- High input impedance 10^{12} Ω

The ECG module and other sensors are connected to the analog input ports, i.e., PA0–PA7 of Atmel microcontroller. These ports PA0–PA7 serve as analog inputs to analog to digital converter.

4 Android-Based Smartphone

Nowadays, Android is the very popular operating system in the smartphones. Its popularity is increasing day by day. It has simple and powerful Java-based development kit and ability to develop on any platform such as Windows, Linux, or Mac. So, the user can develop an android application [3] according to its requirements. Android software development kit (ADK) is used to develop Android applications. Eclipse is the officially supported integrated development environment (IDE) (Fig. 2).

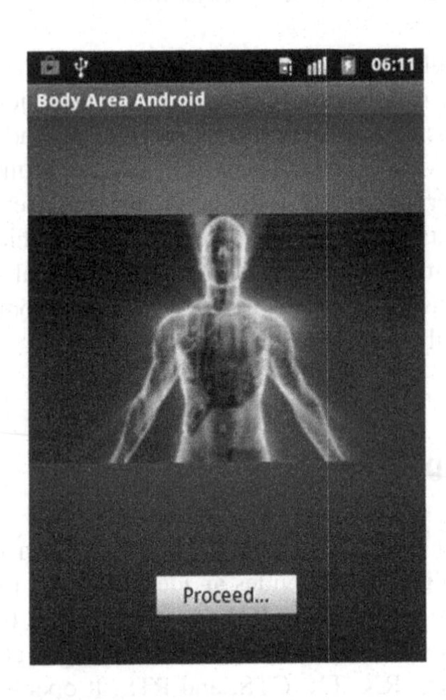

Fig. 2 Android prototype application

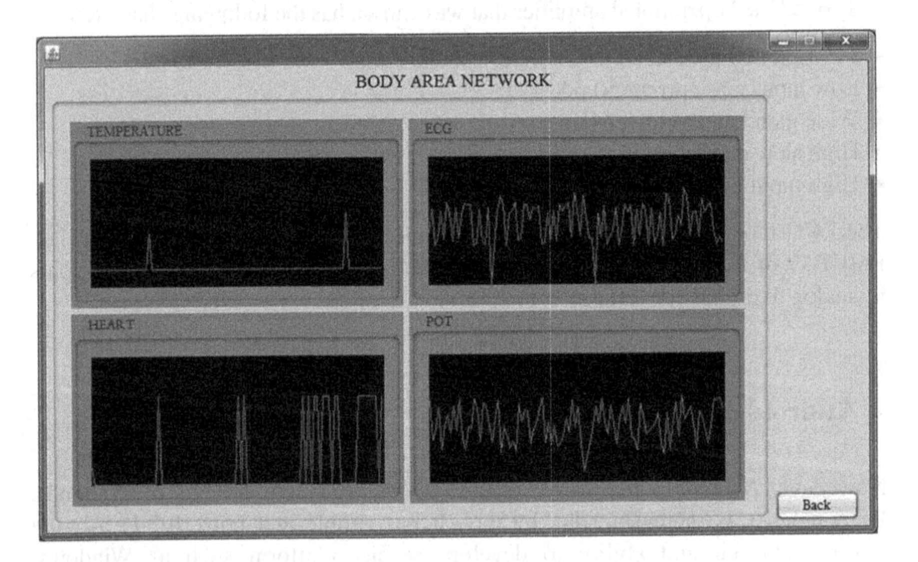

Fig. 3 Main screen at local user

5 Experimental Results

When the initial setting activity parameters are filled, it opens the main screen of the android application. On this main screen of the android application, three physiological parameters can be monitored such as a temperature sensor, heart rate sensor, and ECG sensor [4]. Figure 3 shows the different physiological parameters of the patient. As the alert system is introduced in the system, whenever the reading of the sensor crosses the threshold value, the system will automatically activate the alert. The alert is of two types, i.e., SMS alert and email alert.

6 Conclusion

Real-time monitoring and evaluation of medical parameters of the patient greatly helps to save their life in critical situations. Whenever the reading of the sensor crosses the threshold value, the system will automatically activate the alert. Security on all levels of the layered system must be further investigated. Certification according to medical safety standards is currently impossible due to the different components used, e.g., the Android operating system.

References

1. Kohno, R., Hamaguchi, K., Li, H.-B., Takizawa, K.: R&D and standardization of body area network (BAN) for medical healthcare. In: IEEE International Conference on ICUWB 2008, vol. 3, pp. 5–8, 10–12 Sept 2008
2. Wagner, M., Kuch, B., Cabrera, C., Enoksson, P., Sieber, A.: Android based body area network for the evaluation of medical parameters. In: International Workshop on Intelligent Solutions in Embedded System, pp. 33–38, May 2012
3. Issac, R., Ajaynath, M.: CUEDETA: A real time heart monitoring system using android smartphone. In: 2012 IEEE, pp. 47–52, June 2012
4. Yuan, B., Herbert, J.: Web-based real-time remote monitoring for pervasive healthcare. In: IEEE International Conference on Pervasive Computing and Communications Workshops (PERCOM Workshops), pp. 625 (2011)

An Efficient Kernelized Fuzzy Possibilistic C-Means for High-Dimensional Data Clustering

B. Shanmugapriya and M. Punithavalli

Abstract Clustering high-dimensional data has been a major concern owing to the intrinsic sparsity of the data points. Several recent research results signify that in case of high-dimensional data, even the notion of proximity or clustering possibly will not be significant. Fuzzy c-means (FCM) and possibilistic c-means (PCM) have the capability to handle the high-dimensional data, whereas FCM is sensitive to noise and PCM requires appropriate initialization to converge to nearly global minimum. Hence, to overcome this issue, a fuzzy possibilistic c-means (FPCM) with symmetry-based distance measure has been proposed which can find out the number of clusters that exist in a dataset. Also, an efficient kernelized fuzzy possibilistic c-means (KFPCM) algorithm has been proposed for effective clustering results. The proposed KFPCM uses a distance measure which is based on the kernel-induced distance measure. FPCM combines the advantages of both FCM and PCM; moreover, the kernel-induced distance measure helps in obtaining better clustering results in case of high-dimensional data. The proposed KFPCM is evaluated using datasets such as Iris, Wine, Lymphography, Lung Cancer, and Diabetes in terms of clustering accuracy, number of iterations, and execution time. The results prove the effectiveness of the proposed KFPCM.

Keywords High-dimensional data clustering · Fuzzy c-means (FCM) · Possibilistic c-means (PCM) · Fuzzy possibilistic c-means (FPCM) · Kernelized fuzzy possibilistic c-means (KFPCM) · Kernel-induced distance measure

B. Shanmugapriya (✉) · M. Punithavalli
Sri Ramakrishna College of Arts and Science for Women, Coimbatore, India
e-mail: bspriya2004@yahoo.co.in

M. Punithavalli
e-mail: mpunitha_srcw@yahoo.co.in

I.K. Sethi (ed.), *Computational Vision and Robotics*, Advances in Intelligent Systems and Computing 332, DOI 10.1007/978-81-322-2196-8_26

1 Introduction

Clustering is to identify classes of similar objects among a set of objects, or more specifically, the partitioning of a dataset into a subset of objects, with the intention that the data present in each subset possibly share certain similar features based on certain defined distance measure. A distance measure (e.g., Euclidean distance) between two points is used to measure the dissimilarity between the corresponding objects [1–3].

In case of high-dimensional data, there are several dimensions which are frequently irrelevant. These irrelevant dimensions cannot be handled properly by the traditional clustering approaches [4]. The following are the challenges for the traditional unsupervised clustering algorithms.

- The presence of irrelevant and noisy parameters can affect the clustering results.
- The high-dimensional data may be sparse
- Moreover, when there is increase in the number of dimensions in a dataset, the distance measures in the traditional clustering approaches become increasingly insignificant.

Several techniques are required to handle the high-dimensional datasets and to effectively represent this high-dimensional data into an interpretable and meaningful information [5, 6]. Global feature transformation techniques (e.g., PCA) preserve to some extent the information from irrelevant attributes, and they may thus be unable to identify clusters that exist in different subspaces [7–9]. Projected clustering has been mainly motivated by seminal research showing that, as the dimensionality increases, the farthest neighbor of a point is expected to be almost as close as its nearest neighbor for a wide range of data distributions and distance functions [7]. Furthermore, data objects may cluster differently in varying subspaces [8, 10, 11].

Beyer et al. [12] have shown that the distance of any two points in a high-dimensional space is almost the same for a large class of common distributions. On the other hand, the widely used distance measures are more meaningful in subsets (i.e., projections) of the high-dimensional space, where the object values are dense [13]. In other words, it is more likely for the data to form dense, feasibility clusters in a high-dimensional subspace [4, 14]. Virtually, all existing projected clustering algorithms (PROCLUS, DOC/FASTDOC, HARP, SSPC, and EPCH) assume the meaningful structure, explicitly or implicitly [15, 16].

In order to handle the uncertainty in clustering, it is necessary to bring in certain "fuzziness" into the formulation of the issue. This is the primary reason for using fuzzy models in clustering. In this paper, a KFPCM has been proposed with the purpose of handling this high-dimensional data. The proposed KFPCM uses a distance measure which is based on the kernel-induced distance measure. Since FPCM combines the advantages of both FCM and PCM, it provides better results.

2 Literature Survey

FCM and FPCM are two mainly used fuzzy partition clustering algorithms that are based on Euclidean distance function, which can only be used to notice spherical structural clusters. On the other hand, FCM's noise sensitivity difficulty and PCM's overlapping cluster difficulty are also familiar. In recent times, there have been large number of attempts to integrate both of them to alleviate these problems and PFCM showed promising outcomes.

Namkoong et al. [17] proposed a modified PFCM by means of regularization to decrease noise sensitivity in PFCM further. Yan and Chen [18] carried out advanced evaluation of PFCM for high-dimensional data and developed an enhanced version of PFCM called Hyperspherical PFCM (HPFCM), in which the cosine similarity measure has been integrated. In earlier studies, there are four improved algorithms developed from previous works, namely FCM-M, FPCM-M, FCM-CM, and FPCM-CM based on unsupervised Mahalanobis distance without any extra prior information. An advanced new algorithm called "fuzzy possibilistic c-means based on complete Mahalanobis distance and separable criterion (FPCM-CMS)" is suggested by Chuang et al. [14]. It can get additional information and advanced precision by taking the extra separate principle than FPCM-CM.

A fuzzy clustering technique was proposed by Wu and Zhou [19] depending on the kernel techniques. This model is called as kernel possibilistic fuzzy C-means model. It is declared that this model is an extension of possibilistic fuzzy c-means model (PFCM) which is better than FCM model. PFCM and FCM almost depend on Euclidean distance, but this model depends on non-Euclidean distance by exploiting kernel techniques. In addition, with kernel techniques, the input data can be mapped completely into a high-dimensional feature space in which the nonlinear pattern now seems to be linear. This model can handle data with the presence of noises or outliers. This model is interesting and provides better results than PFCM.

3 Methodology

3.1 Fuzzy C-Means (FCM)

FCM is an iterative clustering approach that produces an optimal c partition by reducing the weighted within group sum of squared error objective function:

$$J_{\text{FCM}}(V,U,X)=\sum_{i=1}^{c}\sum_{j=1}^{n}u_{ij}^{m}d^{2}(x_j,v_i),\quad 1<m<+\infty \tag{1}$$

where $X=\{x_1,x_2,\ldots,x_n\}\subseteq R^p$ is the dataset in the p-dimensional vector space, p represents the number of data items, and c represents the number of clusters with

$2 \leq c \leq n-1$. $V = \{v_1, v_2, \ldots, v_c\}$ is the c centers or prototypes of the clusters, v_i is the p-dimension center of the cluster i, and $d^2(x_j, v_i)$ is a distance between object x_j and cluster center v_i. $U = \{u_{ij}\}$ symbolizes a fuzzy partition matrix with $u_{ij} = u_i(x_j)$ is the degree of membership of x_j in the i-th cluster, x_j is the j-th of p-dimensional measured data. The fuzzy partition matrix satisfies:

$$0 \prec \sum_{j=1}^{n} u_{ij} \prec n, \quad \forall i \in \{1, \ldots, c\} \tag{2}$$

$$\sum_{i=1}^{c} u_{ij} = 1, \quad \forall j \in \{1, \ldots, n\} \tag{3}$$

The parameter m is a weighting exponent on every fuzzy membership and discovers the amount of fuzziness of the resultant classification [20]; it is a predetermined number larger than one. The objective function J_{FCM} can be reduced under the parameter of U. In particular, taking J_{FCM} in accordance with the u_{ij} and v_i and zeroing them in the same way, it is essential but not adequate conditions for J_{FCM} to be at its local extrema. This is given as:

$$\mu_{ij} = \left[\sum_{k=1}^{c} \left(\frac{\mathrm{d}(x_j, v_i)}{\mathrm{d}(x_j, v_k)} \right)^{2/(m-1)} \right]^{-1}, \quad 1 \leq i \leq c,\ 1 \leq j \leq n. \tag{4}$$

$$v_i = \frac{\sum_{k=1}^{n} u_{ik}^{m} x_k}{\sum_{k=1}^{n} u_{ik}^{m}}, \quad 1 \leq i \leq c \tag{5}$$

Eventhough FCM is a very useful clustering approach, its memberships will not always communicate well to the degree of belonging of the data and possibly will be inaccurate in a noisy environment, because the real data inevitably engages with some noises [21].

3.2 *Modified Projected K-Means Clustering Algorithm*

Projected clustering is related to subspace clustering [22] in that both detect clusters of objects that exist in subspaces of a dataset. k-means algorithm with projected clustering deals with clustering for large dataset and it can be used to rectify complex problems.

3.2.1 Setting Virtual Values

The irrelevant dimensions are given as follows. Suppose already there is a k-partition, then for every cluster X_i and each dimension D_j, compute

$$\sum\nolimits_{X \in X_i} (1 - \lambda_j) d_l^2(x_j, c_{ij}) + \lambda_j d_{nl}^2(x_j, c_{ij}) \tag{6}$$

and

$$\sum\nolimits_{X \in X_i} \left(1 - \lambda_j'\right) d_l'^2\left(x_j', c_{ij}'\right) + \lambda_j' d_{nl}'^2\left(x_j', c_{ij}'\right) \quad (7)$$

where x_j denotes the value of data point X on dimension j, c_{ij} represents the mean of data points in cluster X_i on dimension D_j, x_j' indicates the virtual value of data point X on dimension j, c_{ij}' denotes the virtual mean of data points in cluster X_i on dimension D_j. It is unnecessary to set virtual values for each data point, only the sum should be calculated. If n_i data points distribute uniformly on dimension D_j in the range of [Min, Max], the sum is:

$$\sum_{X \in X_i} \left(1 - \lambda_j'\right) d_l'^2\left(x_j', c_{ij}'\right) + \lambda_j' d_{nl}'^2\left(x_j', c_{ij}'\right) = \begin{cases} 2\left(\frac{\max - \min}{n_i - 1}\right)^2 \left[0.5^2 + 1.5^2 + \cdots + \left(\frac{n_i - 1}{2}\right)^2\right], \\ \quad \text{when } n_i \text{ is an even number} \\ 2\left(\frac{\max - \min}{n_i - 1}\right)^2 \left[1^2 + 2^2 + \cdots + \left(\frac{n_i - 1}{2}\right)^2\right], \\ \quad \text{when } n_i \text{ is an odd number} \end{cases} \quad (8)$$

If $\sum\nolimits_{X \in X_i} (1 - \lambda_j) d_l^2 (x_j, c_{ij}) + \lambda_j d_{nl}^2 (x_j, c_{ij}) < \sum\nolimits_{X \in X_i} \left(1 - \lambda_j'\right) d_l^2 \left(x_j', c_{ij}'\right) + \lambda_j' d_{nl}^2 \left(x_j', c_{ij}'\right)$, dimension D_j is relevant to cluster X_i, and if $\sum\nolimits_{X \in X_i} (1 - \lambda_j) d_l^2 (x_j, c_{ij}) + \lambda_j d_{nl}^2 (x_j, c_{ij}) = \sum\nolimits_{X \in X_i} \left(1 - \lambda_j'\right) d_l^2 \left(x_j', c_{ij}'\right) + \lambda_j' d_{nl}^2 \left(x_j', c_{ij}'\right)$, dimension D_j is irrelevant to cluster X_i. However, in real dataset, the allocation of the projections of data points on irrelevant dimension will be more complicated than uniform distribution. A parameter ε is introduced to solve this difficulty. If $\sum\nolimits_{X \in X_i} (1 - \lambda_j) d_l^2 (x_j, c_{ij}) + \lambda_j d_{nl}^2 (x_j, c_{ij}) < \varepsilon . \sum\nolimits_{X \in X_i} \left(1 - \lambda_j'\right) d_l^2 \left(x_j', c_{ij}'\right) + \lambda_j' d_{nl}^2 \left(x_j', c_{ij}'\right)$, dimension D_j is appropriate to cluster X_i, and if $\sum\nolimits_{X \in X_i} (1 - \lambda_j) d_l^2 (x_j, c_{ij}) + \lambda_j d_{nl}^2 (x_j, c_{ij}) \geq \varepsilon . \sum\nolimits_{X \in X_i} \left(1 - \lambda_j'\right) d_l^2 \left(x_j', c_{ij}'\right) + \lambda_j' d_{nl}^2 \left(x_j', c_{ij}'\right)$, dimension D_j is irrelevant to cluster X_i. The range of ε is (0, 1). It is simple to select ε. When the sum of squared-errors of virtual values is set, an appropriate ε will guarantee that $E(W, C)$ satisfies the two requirements mentioned previously.

3.2.2 The Modified Projected K-Means Algorithm

Similar to conventional k-means clustering approach, the proposed algorithm arbitrarily selects k data points in X as the primary cluster centers. Each cluster center C_i is related to a vector W_i whose components equal to one. Then, the proposed algorithm repeats the following two steps to optimize the objective function $E(W, C)$.

1. Allocate each data point in X to the nearest cluster. This results in a k-partition. The distance between a data point X and a cluster X_i is given as below,

$$\mathrm{dis}(X, X_i) = \sqrt{\sum_{j=1}^{d} w_{ij}\left[(1-\lambda_j)d_l^2(x_j, c_{ij}) + \lambda_j d_{nl}^2(x_j, c_{ij})\right] / \sum_{j=1}^{d} w_{ij}} \quad (9)$$

2. Update C_i and W_i for $X_i, 1 \le i \le k$. $c_{ij} = \sum_{X \in X_i} x_j / |X_i|, 1 \le j \le d$, in which $|X_i|$ is the number of data points in X_i.

$$w_{ij} = 1, \text{ when } \sum_{X \in X_i} (1-\lambda_j)d_l^2(x_j, c_{ij}) + \lambda_j d_{nl}^2(x_j, c_{ij}) < \varepsilon . \sum_{X \in X_i} \left(1-\lambda_j'\right)d_l^2\left(x_j', c_{ij}'\right) + \lambda_j' d_{nl}^2\left(x_j', c_{ij}'\right)$$

$$w_{ij} = 0, \text{ when } \sum_{X \in X_i} (1-\lambda_j)d_l^2(x_j, c_{ij}) + \lambda_j d_{nl}^2(x_j, c_{ij}) \ge \varepsilon . \sum_{X \in X_i} \left(1-\lambda_j'\right)d_l^2\left(x_j', c_{ij}'\right) + \lambda_j' d_{nl}^2\left(x_j', c_{ij}'\right)$$

After getting a k-partition, initially the means of the clusters are computed. Subsequently, the weight vector of each cluster is found based on these means. The above mentioned two steps are continuously repeated until the partition does not undergo any change.

3.3 FPCM with Symmetry-Based Distance Measure (FPCM-SDM)

The fitness of FPCM is calculated using the FSym-index. Let K cluster centers be represented by $\bar{c_i}$ in which, $1 \le i \le K$ and $U(X) = [u_{ij}]_{k \times n}$ is a partition matrix for the data [2]. Then, FSym-index is defined as follows:

$$\mathrm{FSym}(K) = \left(\frac{1}{K} \times \frac{1}{E_K} \times D_K\right) \quad (10)$$

where K is the number of clusters. Here,

$$E_K = \sum_{i=1}^{K} E_i \quad (11)$$

where

$$E_i = \sum_{j=1}^{n} \left(u_{ij} \times d_{ps}\left(\bar{x_j}, \bar{c_i}\right)\right) \quad (12)$$

and

$$D_K = \max_{i,j=1} K \left\| \bar{c_i} - \bar{c_j} \right\| \tag{13}$$

D_K is the highest Euclidean distance between two cluster centers among all centers.

The proposed point symmetry-based distance $d_{ps}(\overline{x}, \overline{c})$ associated with point $\overline{x}$ in accordance with a center $\overline{c}$ is defined as follows: Consider a point be $\overline{x}$. The symmetrical point of $\overline{x}$ associated with a particular center $\overline{c}$ is $(2 \times \overline{c} - \overline{x})$. Consider this as $\overline{x}^*$. Also, consider the first and second unique nearest neighbors of $\overline{x}^*$ be at Euclidean distances of d_1 and d_2, respectively. $d_{ps}\left(\bar{x_j}, \bar{c_i}\right)$ is evaluated using the following equation,

$$d_{ps}\left(\bar{x}, \bar{c}\right) = \frac{(d_1 + d_2)}{2} \times d_e(\bar{x}, \bar{c}) \tag{14}$$

where $d_e(\overline{x}, \overline{c})$ is the Euclidean distance between the point $\overline{x}$ and $\overline{c}$.

The major purpose is to increase the value of FSym-index with the aim of obtaining the actual number of clusters and to accomplish proper clustering. As given in 10, FSym is a composition of three factors, these are $1/K, 1/E_K$, and D_K. The initial factor increases when K increases; as FSym is required to be larger for optimal clustering, as a result it will prefer to decrease the value of K. The second factor is the inside cluster total symmetrical distance. For clusters which have better symmetrical structure, E_i value is less. This, consecutively, signifies that construction of more number of clusters, which are symmetrical in shape, would be encouraged. At last the third factor, D_K, measuring the highest separation between a pair of clusters, increases with the value of K. It is to be noted that all these three factors are complementary in nature, and as a result, they are estimated to compete and make stability to each other significantly for determining the proper partitioning to get the most accurate results.

3.4 FPCM Based on Kernel-Induced Distance (KFPCM)

Consider $\Phi : x \in X \subseteq R^d \mapsto \Phi(x) \in F \subseteq R^H (d \ll H)$ is a nonlinear transformation into a higher (probably infinite) dimensional feature space F. This section discusses the implementation of using the kernel methods. When $x = [x_1, x_2]^T$ and $\Phi(x) = \left[x_1^2, \sqrt{2}x_1, x_2, x_2^2\right]^T$, where x_i is the i-th component of vector x. Subsequently, the inner product between $\Phi(x)$ and $\Phi(y)$ in the feature space F are as follows: $\Phi(x)^T \Phi(y) = \left[x_1^2, \sqrt{2}x_1, x_2, x_2^2\right]^T \left[y_1^2, \sqrt{2}y, y_2, y_2^2\right]^T = \left(x^T y\right)^2 = K(x, y)$. Consequently, to calculate the inner products in F, kernel representation $K(x, y)$ is utilized, without unambiguously using transformation or mapping Φ (therefore, overcoming curse of dimensionality). It is a direct outcome from [23] each linear algorithm that only utilizes inner products can be simply extended to a nonlinear version only through the kernels satisfying the Mercer's conditions [23]. The following equation gives the classic radial basis function (RBF) and polynomial kernels:

$$K(x,y) = \exp\left(\frac{-\left(\sum_{i=1}^{d} |x_i - y_i|^a\right)^b}{\sigma^2}\right) \tag{15}$$

where d represents the dimension of vector x; $a \geq 0$; $1 \leq b \leq 2$. It is clear that $K(x,x) = 1$ for all x and the above RBF kernels, and a polynomial with degree of P.

$$K(x,y) = \left(x^T y + 1\right)^P \tag{16}$$

Based on the above formulations, the kernelized version of the FPCM algorithm can be constructed and transform its objective function with the mapping Φ as follows:

$$J_{\text{KFPCM}}^{\Phi} = \sum_{i=1}^{c}\sum_{j=1}^{n}(u_{ij}^m + t_{ij}^n)\left\|\Phi(x_j) - \Phi(v_i)\right\|^2 \tag{17}$$

Now, with the help of the kernel substitution,

$$\begin{aligned} \|\Phi(x_k) - \Phi(v_i)\|^2 &= (\Phi(x_k) - \Phi(v_i))^2(\Phi(x_k) - \Phi(v_i)) \\ &= \Phi(x_k)^{\mathrm{T}}\Phi(x_k) - \Phi(v_i)^{\mathrm{T}}\Phi(x_k) \\ &\quad - \Phi(x_k)^{\mathrm{T}}\Phi(v_i) + \Phi(v_i)^{\mathrm{T}}\Phi(v_i) \\ &= K(x_k, x_k)K(v_i, v_i) - 2K(x_k, v_i) \end{aligned} \tag{18}$$

By this manner, a novel class of non-Euclidean distance measures in original data space is acquired. Noticeably, several kernels will stimulate different measures for the original space, which generates a new family of clustering approaches. In particular, when the kernel function K(x, y) is considered as the RBF in (15), (18) can be made simpler to $2(1 - K(x_k, v_i))$. In order to reduce the manipulation below and robustness considered, only the Gaussian RBF (GRBF) kernel with a=1 and b=1 in (15) then (17) can be modified as

$$J_{\text{KFPCM}}^{\Phi} = 2\sum_{i=1}^{c}\sum_{j=1}^{n}(u_{ij}^m + t_{ij}^n)\left(1 - K\left(x_j, v_i\right)\right) \tag{19}$$

By an optimization way similar to the J_{KFPCM}^{Φ}, can be reduced under the constraint of U. In particular, consider its first derivatives with respect to u_{ij} and v_i, and zero them, correspondingly, two necessary but not adequate conditions for J_{KFPCM}^{Φ} to be at local minimum will be obtained as

$$u_{ij} = \frac{\left(1 - K\left(x_j, v_i\right)\right)^{-\frac{1}{(\mathrm{m}-1)}}}{\sum_{\mathrm{j}=1}^{\mathrm{c}}\left(1 - K\left(x_j, v_i\right)\right)^{-\frac{1}{(\mathrm{m}-1)}}} \tag{20}$$

$$v_i = \frac{\sum_{j=1}^{n} u_{ij}^{m} K(x_j, v_i) x_j}{\sum_{j=1}^{n} u_{ij}^{m} K(x_j, v_i) x_j} \quad (21)$$

It is obvious that the obtained centroids or prototypes $\{v_i\}$ still lie in the original space and not in the transformed higher dimensional feature space; as a result, the computational simplicity is still retained. Moreover, it is revealed that the KFPCMs resulted from (17) to (19) are robust to outliers and noise based on the Huber's robust statistics [8, 15].

4 Experimental Results

In order to evaluate the proposed KFPCM against the FCM, modified projected k-means clustering and FPCM-SDM algorithms, experiments were carried out using Lung Cancer, Lymphography, Iris, Wine, and Diabetes datasets.

The performance of the proposed algorithm is evaluated based on the following parameters:

- clustering accuracy
- number of iterations, and
- execution time

4.1 Clustering Accuracy

From the figure, it can be observed that for all the datasets, the accuracy of clustering results using FCM and FPCM is very low than that of the proposed KFPCM which obtains 98.3 % accuracy in Lung Cancer dataset, 99.7 % in Lymphography dataset, 99.4 % in Iris dataset, 98.7 % in Wine dataset, and 97.9 % in Diabetes datasets. It is clear from the Fig. 1 that proposed KFPCM is better than the other three approaches.

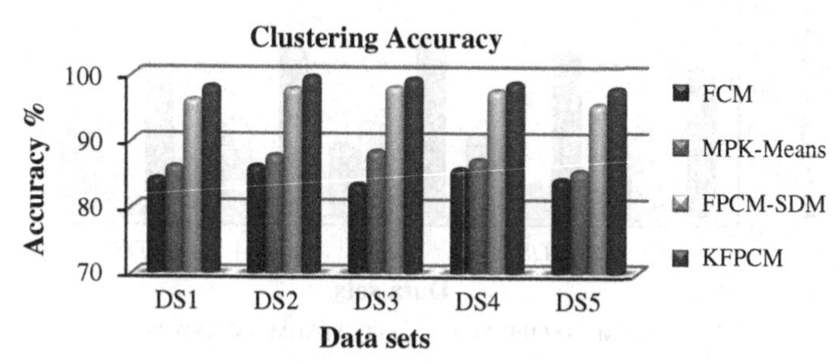

Fig. 1 Comparison of clustering accuracy

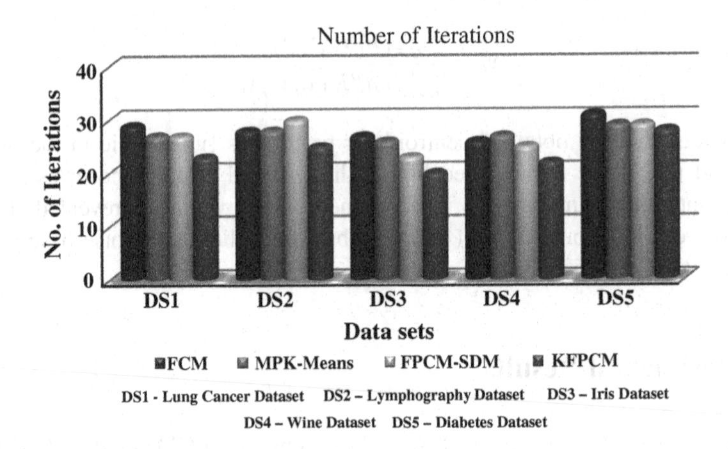

Fig. 2 Comparison of number of iterations

4.2 Number of Iterations

From the Fig. 2, it can be observed that the number of iterations required for execution using the proposed KFPCM on Lung Cancer dataset is 23, for Lymphography dataset is 25, for Iris dataset is 20, for Wine dataset is 22, and for Diabetes dataset is 28, whereas more iterations is required for other two clustering techniques for the desired result. From Fig. 2, it is observable that the proposed algorithm utilizes only very less iterations than the other three algorithms.

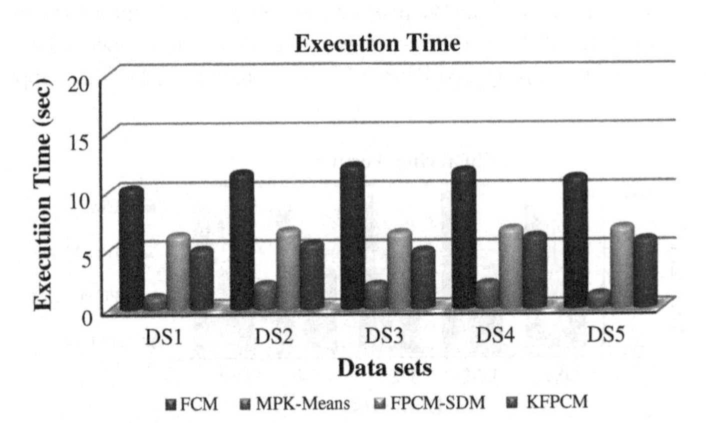

Fig. 3 Comparison of execution time

4.3 Execution Time

From the Fig. 3, it can be observed that the time required for execution using the proposed KFPCM on Lung Cancer dataset is 5.1 s, for Lymphography dataset is 5.6 s, for Iris dataset is 5 s, for Wine dataset is 6.2 s, and for Diabetes dataset is 5.9 s, whereas more time is required for other two clustering techniques for execution. From Fig. 3, it is obvious that the proposed algorithm takes very less execution time than the other three algorithms.

5 Conclusion

Projected clustering is aggravated by datasets with a large number of attributes or with irrelevant attributes. Existing projected clustering algorithms often fail to cluster the high-dimensional data. In recent times, the familiar "kernel techniques" has been widely applied in unsupervised clustering. In this paper, a novel kernelized fuzzy possibilistic c-means (KFPCM) has been proposed based on the kernel-induced distance measure. It is clear from the experimental results that the proposed KFPCM clustering approach provides better clustering results than the FCM, modified projected k-means clustering, and FPCM-SDM approaches with more accuracy in less number of iterations.

References

1. Yip, K.Y., Cheung, D.W., Ng, M.K.: On discovery of extremely low-dimensional clusters using semi-supervised projected clustering. In: ICDE (2005)
2. Moise, G., et al.: P3C: a robust projected clustering algorithm. Department of Computing Science, University of Alberta
3. Aggarwal, C.C., et al.: A framework for projected clustering of high dimensional data streams. In: Proceedings of the 30th VLDB Conference, Toronto, Canada (2004)
4. Papadopoulos, D., Gunopulos, D., Ma, S.: Subspace clustering of high dimensional data. In: SIAM (2004)
5. Kohane, I.S., Kho, A., Butte, A.J.: Microarrays for an Integrative Genomics. MIT Press, Massachusetts (2002)
6. Raychaudhuri, S., Sutphin, P.D., Chang, J.T., Altman, R.B.: Basic microarray analysis: grouping and feature reduction. Trends Biotechnol. **19**(5), 189–193 (2001)
7. Parsons, L., Haque, E., Liu, H.: Subspace clustering for high dimensional data: a review. SIGKDD Explor. Newsl. **6**(1), 90–105 (2004)
8. Havens, T.C., Chitta, R., Jain, A.K., Jin, R.: Speedup of Fuzzy and possibilistic kernel C-means for large-scale clustering. Department of Computer Science and Engineering, Michigan State University, East Lansing
9. Günnemann, S., et al.: Subspace clustering for indexing high dimensional data: a main memory index based on local reductions and individual multi-representations. In: Proceedings of the 14th International Conference on Extending Database Technology, EDBT 2011, Uppsala, Sweden, 22–24 Mar 2011

10. Zhang, D.-Q., Chen, S.-C.: Kernel-based fuzzy and possibilistic C-means clustering. Nanjing University of Aeronautics and Astronautics, Nanjing
11. Vanisri, D., Loganathan, C.: An efficient fuzzy possibilistic C means with penalized and compensated constraints. Glob. J. Comput. Sci. Technol. **11**(3), (2011)
12. Beyer, K.S., Goldstein, J., Ramakrishnan, R., Shaft, U.: When is "nearest neighbor" meaningful? In: Proceeding of the 7th International Conference on Database Theory (ICDT '99) 1999
13. Hinneburg, A., Aggarwal, C.C., Keim, D.A.: What is the Nearest Neighbor in high dimensional spaces? In: Proceedings of International Conference on Very Large Data Bases (VLDB '00) 2000
14. Chu, Y.H., Huang, J.W., Chuang, K.T., Yang, D.N., Chen, M.S.: Density conscious subspace clustering for high-dimensional data. IEEE Trans. Knowl. Data Eng. **22**(1), (2010)
15. Frigui, H.: Simultaneous clustering and feature discrimination with applications. In: Advances in Fuzzy Clustering and Feature Discrimination with Applications, pp. 285–312. Wiley, New York (2007)
16. Sledge, I., Havens, T., Bezdek, J., Keller, J.: Relational duals of cluster validity functions for the C-means family. IEEE Trans. Fuzzy Syst. **18**(6), 1160–1170 (2010)
17. Namkoong, Y., Heo, G., Woo, Y.W. : An extension of possibilistic fuzzy C-means with regularization. In: IEEE International Conference on Fuzzy Systems (FUZZ), pp. 1–6 (2010)
18. Yan, Y., Chen, L.: Hyperspherical possibilistic fuzzy c-means for High dimensional data clustering. In: 7th International Conference on Information, Communications and Signal Processing (2009)
19. Wu, W.H., Zhou, J.J.: Possibilistic fuzzy c-means clustering model using kernel methods. Comput. Intell. Model. Control Autom. **2**, 465–470 (2005)
20. Sun, Y., Liu, G., Xu, K.: A k-means-based projected clustering algorithm. In: Third International Joint Conference on Computational Science and Optimization (CSO), vol. 1, pp. 466–470 (2010)
21. Olive, D.J.: Applied Robust Statistics. Carbondale, 62901-4408 (2008)
22. Agrawal, R., Gehrkem, J., Gunopulos, D., Raghavan, P.: Automatic subspace clustering of high dimensional data for data mining applications. In: Haas, L., Tiwary, A. (eds) Proceedings of the ACM SIGMOD International Conference on Management of Data, pp. 94–105. Seattle ,WA (1998)
23. Jia, K., He, M., Cheng, T.: A new similarity measure based robust possibilistic C-means clustering algorithm. Lect. Notes Comput. Sci. **6988**, 335–342 (2011)

Author Index

I.K. Sethi (ed.), *Computational Vision and Robotics*, Advances in Intelligent Systems and Computing 332, DOI 10.1007/978-81-322-2196-8